Engines

Engines

The **Inner Workings** of **Machines** That **Move the World**

Theodore Gray

Bestselling author of *The Elements* and *How Things Work*
Photographs by Nick Mann

BLACK DOG
& LEVENTHAL
PUBLISHERS
NEW YORK

Black Dog & Leventhal Publishers
Hachette Book Group
1290 Avenue of the Americas
New York, NY 10104

www.hachettebookgroup.com
www.blackdogandleventhal.com

First Edition: October 2022

Black Dog & Leventhal Publishers is an imprint of Perseus Books, LLC,
a subsidiary of Hachette Book Group, Inc. The Black Dog & Leventhal Publishers
name and logo are trademarks of Hachette Book Group, Inc.

The publisher is not responsible for websites (or their content) that are not owned by the publisher.

Additional copyright/credits information is on page 234.

Print book interior design by Matthew Riley Cokeley & Kevin Beard

Library of Congress Cataloging-in-Publication Data

Names: Gray, Theodore W., author. | Mann, Nick, photographer.
Title: Engines: the inner workings of machines that move the world /
Theodore Gray; photographs by Nick Mann.
Description: First edition. | New York: Black Dog & Leventhal Publishers,
Hachette Book Group, 2022. | Includes index. | Summary: "International best-selling author Theodore Gray, whose Elements Trilogy has sold more than 1.5 million copies, now turns his irrepressible curiosity, wonderful wit, and inimitable photography to the world of engines in this visual exploration of everything from the first steam engines to giant turbines to tiny electrical engines"—Provided by publisher.
Identifiers: LCCN 2021026205 (print) | LCCN 2021026206 (ebook) | ISBN 9780762498345 (hardcover) | ISBN 9780762498321 (ebook)
Subjects: LCSH: Engines—Popular works. | Motors—Popular works.
Classification: LCC TJ250 .G69 2022 (print) | LCC TJ250 (ebook) | DDC 621.4—dc23
LC record available at https://lccn.loc.gov/2021026205
LC ebook record available at https://lccn.loc.gov/2021026206ISBNs: 978-0-7624-9834-5 (hardcover), 978-0-7624-9832-1 (ebook)

ISBNs: 9780762498345 (hardcover), 9780762498321 (ebook)

Printed in China

1010

10 9 8 7 6 5 4 3 2 1

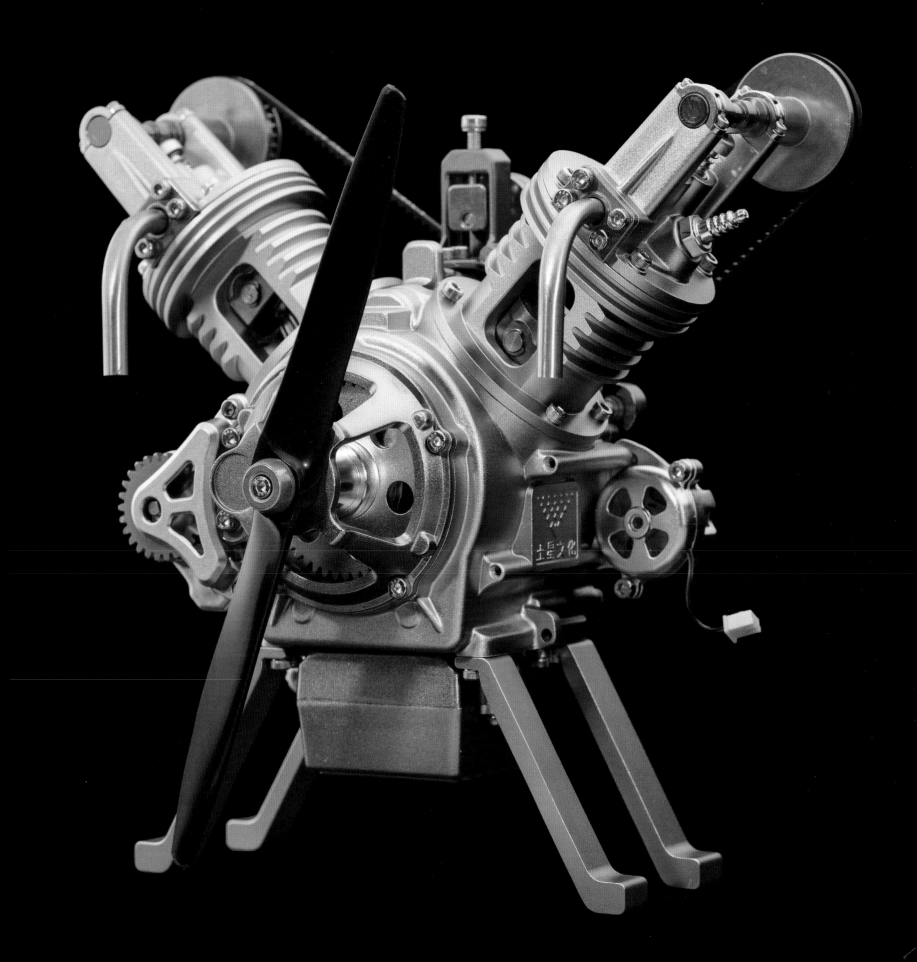

Contents

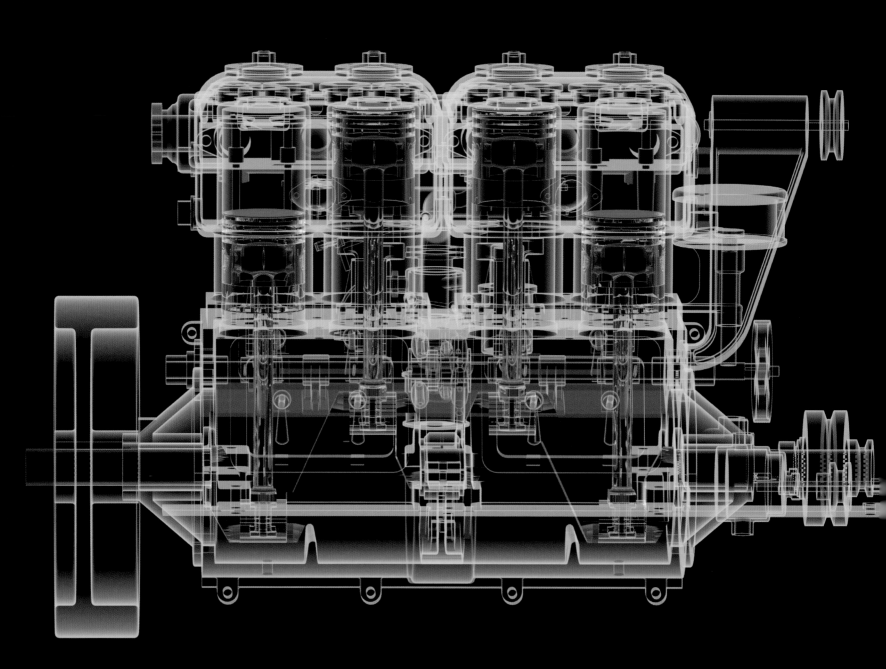

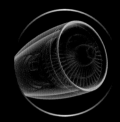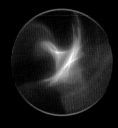

▷ Steam engines are old—they were the first real engines—but they are also new. A lot of our electricity is still produced by steam turbine engines powered by nuclear reactors, coal, or natural gas.

Engines: How They Work and Why We Love Them

ENGINES HAVE FREED us from the limits of human and animal labor. From a diesel engine the size of a small apartment building, to a piezoelectric motor so small a surgeon was able to thread it through the arteries in my heart, engines push us far beyond the limits of mere human ability. No human is as strong as even a modest electric motor. No human can move as fast, as accurately, or as tirelessly as the hundreds of motors we encounter in daily life. Yet they are so commonplace we take them for granted. Some we even treat as disposable, throwing them away when the batteries run out!

It's impossible to overstate how dependent we have become on engines and the power that flows through them. On average each of us has the equivalent of about a dozen horses—about 10 kilowatts of average continuous power use—working for us twenty-four hours a day, seven days a week. Some is delivered directly in the form of heat (for example, by burning oil or gas), but the rest is fed into or through one form of engine or another.

No matter the type, engines operate by converting energy from some available source—electricity, sunlight, wind, gasoline, natural gas, oil, geothermal heat, and so on—into forceful mechanical movement, usually rotational movement.

In this book we will explore the huge and tiny, simple and complex, fast and slow, noisy and silent, dirty and clean, but always fascinating world of engines. We will look deep inside them to learn how they work, and we'll look outside and around them to appreciate how they have changed the world.

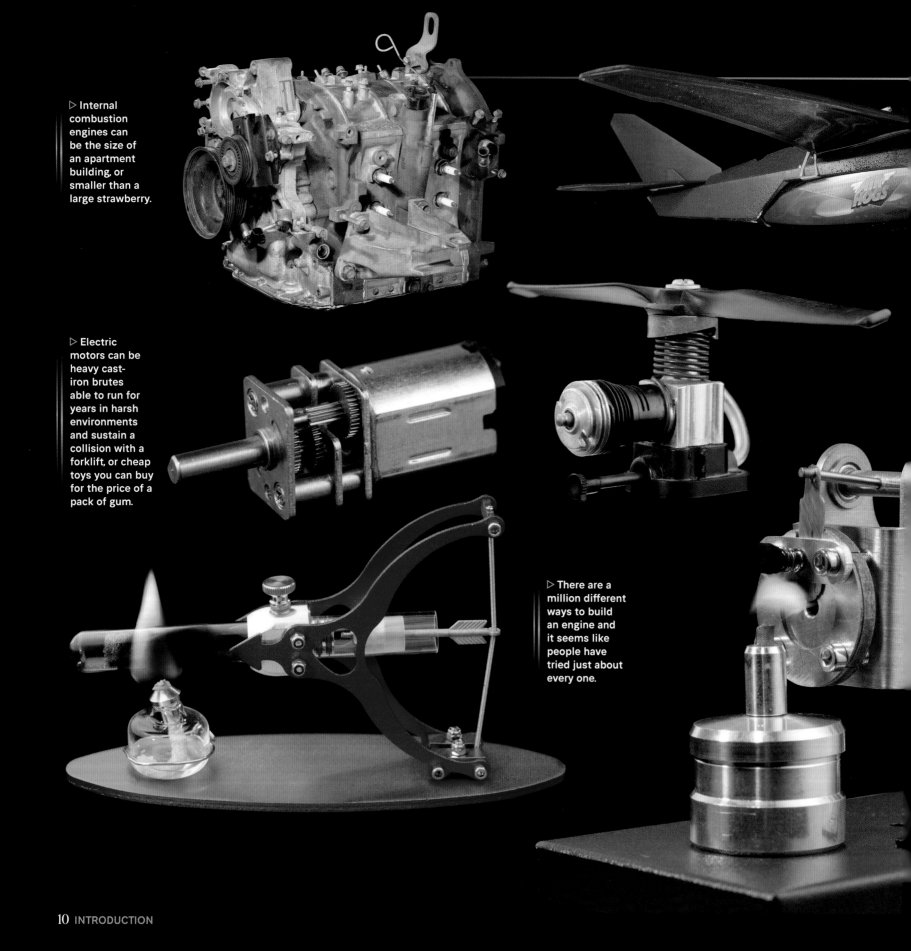

▷ Internal combustion engines can be the size of an apartment building, or smaller than a large strawberry.

▷ Electric motors can be heavy cast-iron brutes able to run for years in harsh environments and sustain a collision with a forklift, or cheap toys you can buy for the price of a pack of gum.

▷ There are a million different ways to build an engine and it seems like people have tried just about every one.

What's the Difference Between an Engine and a Motor?

THERE ISN'T ONE. Although we typically say "car engine" or "electric motor," for all practical purposes these words are interchangeable, and I can prove it to you in one sentence: *Motor* cars have *engines*. Case closed.

▷ Electric motor or electric engine?

▷ This sign is prohibiting cars and other motorized vehicles ... *that have engines.*

▷ Finding examples of gasoline engines being referred to as motors is easy, but here's a more elusive example of an electric motor being referred to as an engine. The maker of this fancy hair trimmer gushes about how it has a "Ferrari-designed engine," by which they mean the electric motor that makes it run.

△ High-torque, brushless Ferrari-designed 7,200–rpm engine

Now that we've got that straight, let's get right to work with the first and best engines, powered by steam.

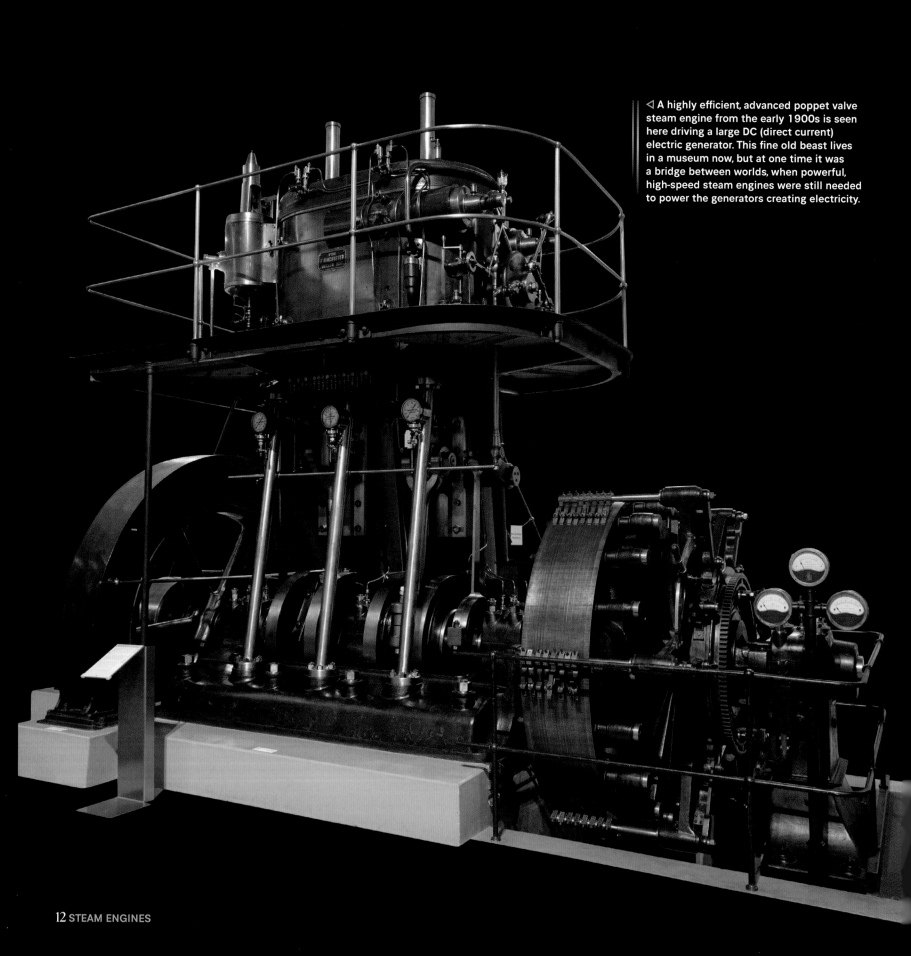

◁ A highly efficient, advanced poppet valve steam engine from the early 1900s is seen here driving a large DC (direct current) electric generator. This fine old beast lives in a museum now, but at one time it was a bridge between worlds, when powerful, high-speed steam engines were still needed to power the generators creating electricity.

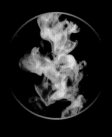

Steam Engines

STEAM ENGINES are the best machines there ever were, or ever will be. They are the purest, most beautiful, elegant, and sublime embodiment of the essence of a machine.

They are huge and oily and they smell just right. Electric motors are too quiet and gasoline engines are too loud, but steam engines are perfect. They huff and they puff and they make music with their valves. They have parts that go around, and they have parts that go back and forth. Other machines might be more practical today, but nothing thrills the heart like a smooth-running steam engine breathing clouds from the fire in its belly.

We owe a lot to the museums and hobbyists who keep the best of them running, and even build new ones from time to time. Without these fans of the Steam Age, these beautiful beasts would be lost to centuries past. Sadly, they just aren't as practical as the newer forms of engines we will encounter in later chapters.

Well, enough romance; let's get on with learning about how these fine engines work. We'll start with the basics, because by understanding the foundations of steam power, we set the stage for understanding the much larger world of internal combustion (gas and diesel) engines to come.

How Steam Engines Work

THE PURPOSE OF an engine is to convert energy into mechanical force or movement. In the case of a steam engine this effort is almost always in the service of making a shaft rotate with some enthusiasm. With a turning shaft you can do all kinds of things, like spin the bit in a drill press or drive the wheels of a car.

As with most engines, what you have to begin with is a *linear force*, that is, a force that pushes along a straight line. So the first job of a steam engine is to turn linear force into circular motion. This is done with a *crank*, which you'll already be familiar with if you've ever ridden or watched someone else ride a bicycle.

When you pedal a bicycle, your legs and feet create a linear force that pushes on the pedals, which are mounted on a crank arm, which in turn is mounted to a big gear. This circuit—legs to crank arm to gear—converts the up-and-down movement of your legs into the rotary motion of the gear. The timing of the pushing and pulling is key. If you keep pushing when you reach the bottom of the stroke, you're not helping the gear keep turning. Instead, you need to quickly switch from pushing down to lifting

up at the exact moment you reach the bottom, and then switch again to pushing down at the top of the rotation. You can do this because you have a sophisticated electronic control system (called a brain) that coordinates the timing of your legs. Some cleverness is needed to get a lump of metal to do the same thing.

The action in a steam engine begins with boiling water in a closed, sealed container or boiler. When the water turns to steam, it expands and creates pressure inside the boiler. A pipe routes the pressurized steam into the cylinder, where it pushes on the end of the piston, forcing it toward (or away from) the crank. The crank turns the flywheel and anything else connected to the crank shaft.

Just as with pedaling a bicycle, timing is key. The steam needs to push on the piston in the right direction at the right time, or the engine won't run. Your muscles move in whatever direction your brain tells them to, and the piston moves in whatever direction the steam pushes it. So in place of a brain, the steam engine has a set of valves that controls when and where steam pressure is applied.

▷ Bicycles are like human-powered steam engines, with legs instead of pistons, a digestive system instead of a boiler, and brains instead of valves.

The boiler supplies energy to the system.

A piston pushes on the crank arm.

The flywheel keeps the movement going through dead spots.

The control valves time the pressure applied by the pistons.

Nick is seen here "letting off steam" by tearing up a tax bill.

▷ A steam engine is a lot like a person riding a bicycle, in metal form. This little model was at one time a cheap toy steam engine, a lot like one I had when I was a kid. Now it's a much more expensive collectible I paid too much for on eBay because I didn't think to save my own. Either way, it was and is a marvelous thing to watch. You fill the boiler with water, pop in a few dry fuel pellets, and a couple minutes later it's puffing away as happy as can be.

A manual speed control valve limits the flow of steam to the piston and valve.

A spring-loaded pressure relief valve opens if the pressure inside the boiler gets dangerously high.

This engine is "letting off steam" through a steam whistle that sounds a lot like someone screaming while tearing up a tax bill.

Water in the boiler is turned to steam under pressure.

This is a pretend "governor" that spins when the engine is running. If it were real it would regulate, or govern, the speed of the engine.

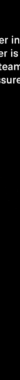

The crank arm translates linear (straight line) motion into circular motion.

The valve, like a simple brain, controls the timing of pressure applied to the piston.

The piston is like a leg pushing on a pedal.

Fuel burning under the boiler heats the water.

The crank arm translates linear motion into circular motion.

The flywheel keeps the movement going through dead spots.

Models of Perfection

TOY STEAM ENGINES have been popular for a long time. Even today you can buy many different types of working toy steam engines. You can still find brand-new cutaway educational models that show exactly how these engines work, despite the fact that no one has actually *needed* to know how a steam engine works since before I was born. To me this feels like the mechanical equivalent of teaching cursive writing, but I'm very glad there are those who still appreciate the magic of these mechanical wonders enough to keep up the tradition.

Toy internal combustion engines are also available, but they are not nearly as popular, probably because they are so much less pleasant to be around when they are running. A steam engine can run gently on just a little steam, air pressure, or canned air. Some of them run so smoothly that you can simply blow into them to make them go. There's something *so* satisfying about a little engine just humming along. But if you want to understand how they work, the cutaway models are what you need.

▷ This clever, fully operational cutaway model does a much better job explaining itself. You can actually see most of the working bits. It runs so easily that you can just blow into the pipe on the left to operate it, except that makes the clear cover over the valves fog up. Better to use compressed air.

The crank arm is visible too.

You can run this steam engine using steam from a little boiler or air from an air compressor.

The flywheel, at least, is easy to see.

It's very hard to see the relationship between the movement of the piston and the valve, which is controlled by this section of the motor.

You can't see the valves, because they're inside this other block of metal.

You can't see the piston at all, because it's inside this block of metal.

△ This chunky, solid cast-iron and brass engine is much nicer than the toy model we were just looking at. It's about 6 inches (15 cm) wide and it runs beautifully and smoothly, fast or slow depending on how much pressure you give it.

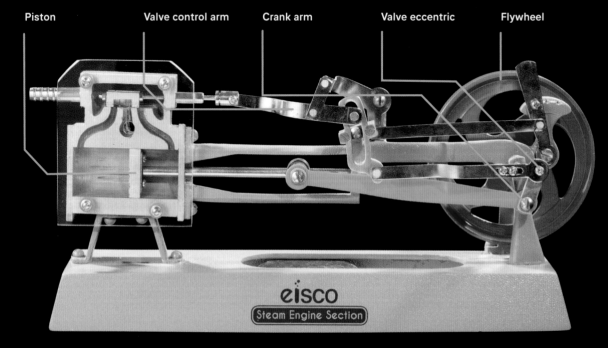

Piston   Valve control arm   Crank arm   Valve eccentric   Flywheel

eisco
Steam Engine Section

Valves make sure the steam gets to the right side of the piston at the right time, but there's also the question of how *much* steam to send. The *governor's* job is to regulate the total amount of steam allowed through, in order to control, or govern, the speed of the engine. How does this work?

Below are two close-up images of a governor. The first image shows the governor in its initial, dormant position, before the engine is running. The second image shows what happens once the engine is up to speed.

The small pulley on the left side of the governor is connected by a belt to the flywheel shaft. When the engine runs, the three balls spin. As the engine increases its speed (as more steam is delivered to the pistons) the balls spin faster. Centrifugal force pushes the balls outward from their center, which forces the arms supporting them to compress, which pushes down on a shaft running through the center of the axle that supports the spinning assembly. This in turn operates a valve that reduces the amount of steam that's delivered to the engine, thus slowing the engine down.

This is a feedback mechanism. The faster the engine turns, the less steam is allowed through. When the engine slows down, the balls begin to return to their original positions, which allows more steam through the system and speeds up the engine once again. If things are working right, these factors balance out and the engine runs at a steady speed. If things are not working properly, the engine "hunts," speeding up and slowing down in an annoying cycle.

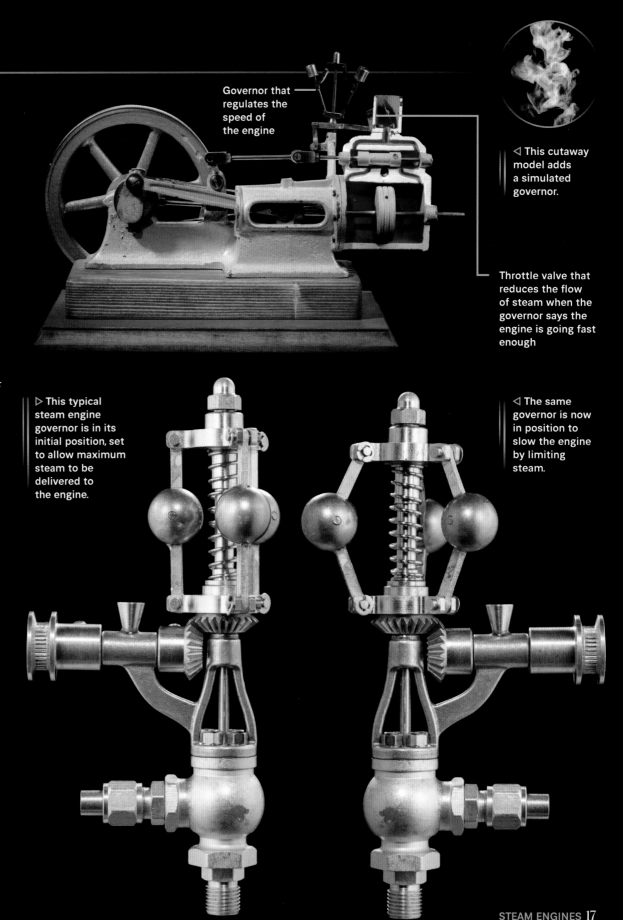

Governor that regulates the speed of the engine

◁ This cutaway model adds a simulated governor.

Throttle valve that reduces the flow of steam when the governor says the engine is going fast enough

▷ This typical steam engine governor is in its initial position, set to allow maximum steam to be delivered to the engine.

◁ The same governor is now in position to slow the engine by limiting steam.

▷ Steam engine models don't have to be elaborate to be beautiful. Sometimes being made entirely of brass is all it takes.

△ Steam engines will run on any kind of compressed gas, not just steam. I discovered a wonderful fact about the tiny brass ones I collected for this book. The inlet ports are exactly the right size to fit snugly into the nozzle of a can of Dust-Off. Stick in the nozzle, pull the trigger, and the engine spins! This is by far the most convenient, expensive, pointless, and environmentally questionable way to run a steam engine.

◁ Steam engines go with anything! Here's one mounted on top of a walking stick, such as might be used by a proper Victorian gentleman. This is not an antique or a one-off built by some hobbyist; it's a commercial product. Clearly there's a market for this kind of accessory.

MADE IN INDIA

▽ This Air Hogs brand model airplane uses a compressed air–powered "steam" engine to make it actually fly. You use a bicycle pump to compress air into what amounts to a plastic soda bottle that makes up the body of the airplane. Instead of just releasing the compressed air to propel the plane, this aircraft is equipped with a propeller that in effect "pushes off" against the surrounding air and allows the plane to fly much farther. The yellow tips on the ends of the blades are heavier than you might think. They act as the flywheel, needed to keep the engine spinning. The engine adds weight, but more than compensates for it by making the translation of stored energy to momentum much more efficient.

▽ The Air Hogs steam engine is a marvel of simplification. A *poppet* valve is pushed open by the end of the cylinder when it reaches the bottom of the stroke. This is by most measures a *terrible* way to design a steam engine, because it means air pressure is applied before the piston reaches a standstill, briefly pushing it back in the wrong direction. But apparently the inertia of the valve means it stays open just a bit longer when the piston is going the way you want it to, resulting in a net force keeping the engine going. The main advantage of this design is that it's extremely cheap to make. The manufacturer even reused a single valve spring to close both the cylinder valve and the filling valve just below it.

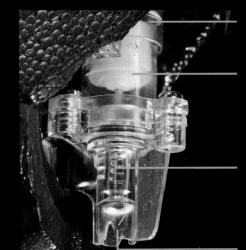

Piston

Cylinder valve that is pushed open (down) when the piston is at the bottom of its stroke

Spring that pushes down on the filling port valve, and up on the cylinder valve

Filling port

◁ This has got to be the cheapest working steam engine model ever—$24 delivered right to my door. And it actually runs! Although steam engines are sometimes held up as examples of early precision machining, the fact is that you don't *need* precision-machined metal components to make a working, even useful, steam engine. This one is made entirely of cheaply molded plastic (except the metal boiler), and some early steam engines were made largely of wood. We'll see in the next chapter that this is *not* the case with internal combustion engines, which absolutely rely on advanced metals and precision machining.

△ Cretors, still in business today as a maker of popcorn poppers, built these entertaining steam-powered popcorn wagons for many decades. The steam engine, mounted right in the middle of the wagon, stirred the popcorn while providing entertainment for the customers. Some of the original wagons are still in use, still with working steam engines, at the Disneyland theme park in Anaheim, California.

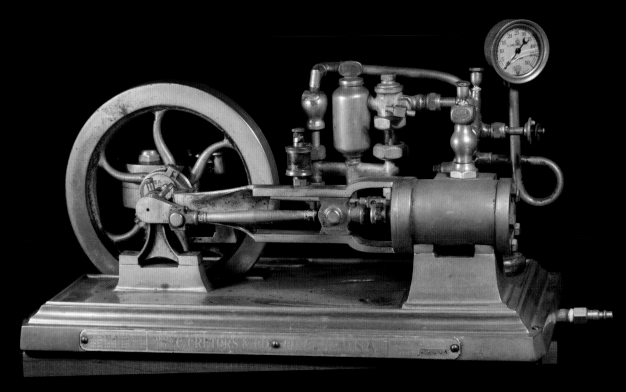

▷ It takes love to make these tiny details just right.

△ Some steam engines blur the line between model and working engine. This one is currently displayed as a model, but it was originally a working engine with a real job to do. What job could such a small engine actually do? Stir popcorn!

STEAM ENGINES CAN be made cheaply, but there's real danger in going too far in that direction: Boilers are not to be taken lightly. The boiler in a toy steam engine is usually just an empty container. Water inside is boiled by heat from below and exits the top as steam. A pressure relief valve is designed to open long before the container bursts. But in a real steam engine the boiler is a much more complicated arrangement of hundreds of metal tubes going back and forth inside what looks like a simple tank. Water flows around these tubes while fire and super-heated gas from the fire flows through them. This greatly increases the amount of heat that can be transferred from the fire to the water, by increasing the surface area where they meet. But it also increases the opportunity for pressure to build up very rapidly.

△ The best views of the insides of steam boilers come when they have a bad day. Needless to say, these explosions are very dangerous: Dozens of people have been killed in single boiler explosions. These disasters continue to occur even in modern times, though usually today it's boilers installed in buildings that have steam or hot water heat. Numerous safety features, such as automatic pressure relief valves, are supposed to prevent explosions, but for every safety feature there is a circumstance in which it can fail.

Anatomy of a Steam Engine

TO REALLY UNDERSTAND how steam engines work, you have to wrap your head around the dynamic interplay of movements between the piston, flywheel, and valves. I spent altogether too long studying the metal models on the previous pages and never really felt like I "got it." What I needed was a model that left out all the nitpicky details required to make an engine actually work and concentrated instead on the fundamentals of time and motion.

It was only after I designed and built this model—which frankly does not look much like a steam engine at all—that I truly understood the simple, fundamental principle that underlies any engine of this type. Once I got the basic idea, all the other models suddenly made sense, and I could move on to design new ones that look and work a lot more like actual steam engines. I hope to show you through these pictures what I learned from the model, though, to be honest, for me at least, it took holding the thing in my hands before it really made sense.

You need to grasp two key ideas: *harmonic* motion and the *phase* of motions relative to each other. But before I explain those two terms, we're going to go through step-by-step, in excruciating detail, the key stages of one complete cycle of my model engine. Bear with me.

▷ This is a highly stylized, simplified model of a steam engine, showing only the key parts.

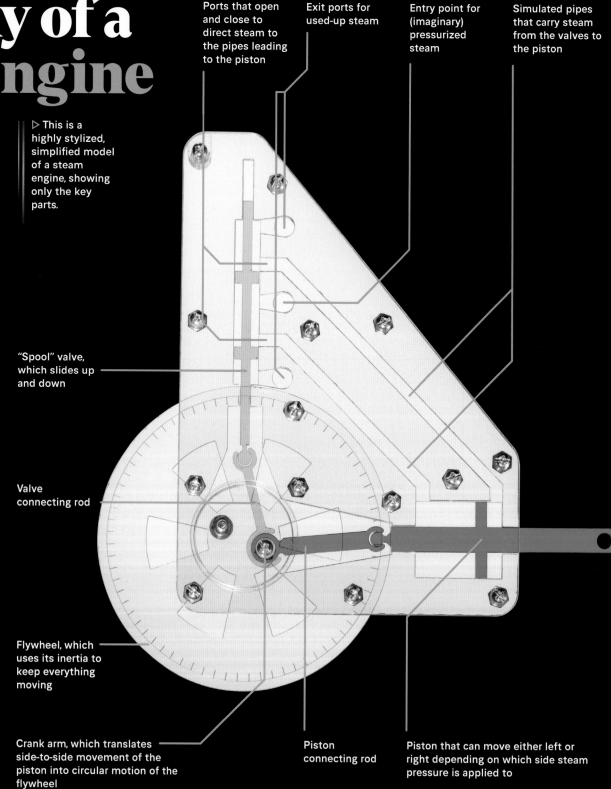

Ports that open and close to direct steam to the pipes leading to the piston

Exit ports for used-up steam

Entry point for (imaginary) pressurized steam

Simulated pipes that carry steam from the valves to the piston

"Spool" valve, which slides up and down

Valve connecting rod

Flywheel, which uses its inertia to keep everything moving

Crank arm, which translates side-to-side movement of the piston into circular motion of the flywheel

Piston connecting rod

Piston that can move either left or right depending on which side steam pressure is applied to

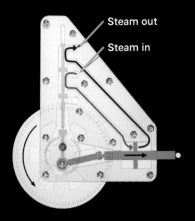

STEP 1 We start with the engine arbitrarily in this position. The piston (green) is traveling toward the right and is about halfway through its stroke. The valve (orange) has opened up a channel to direct high-pressure steam (shown with a red arrow) through the lower pipe to the left side of the piston. At this point the valve is barely moving, and in fact comes to rest for a moment as it changes direction from moving down to moving up.

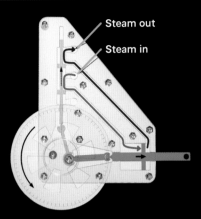

STEP 2 A moment later the piston has almost reached the right side of the cylinder. It's still moving, but slowly. The valve, on the other hand, is moving fast upward. Think about the movements of the piston and valve like your legs when pedaling a bicycle. They go faster as they move through the stroke, but they slow down to a momentary dead stop when reversing direction at the top and bottom of the stroke. The valve and piston both follow this same choreography, but one-quarter turn out of sync: when the piston is moving fastest, the valve is pausing, then reversing direction; and vice versa.

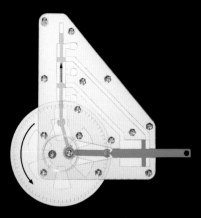

STEP 3 A tiny bit later the piston stops and reverses direction, and the valve has briefly cut off all steam flow. (Notice how the thicker sections of the valve are blocking both of the steam pipes.) At this point the valve is moving at its fastest pace, which means it won't stay shut off for long.

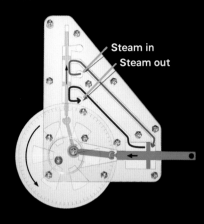

STEP 4 A split second later the valve has moved far enough to open a new channel for steam to flow. Instead of connecting the steam inlet with the lower pipe, now the inlet is connected to the upper pipe. So instead of pushing on the left side of the piston, steam is now directed to the right side of the piston, which reverses the direction of its movement.

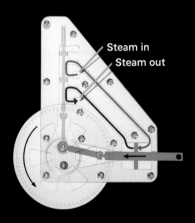

STEP 5 Pushed by steam from the right, the piston is now traveling to the left. Here it's in the middle of its stroke, traveling as fast as it can. The valve is nearly stopped as it begins to change direction. Notice again the complementary nature of the two movements: when one is going fast, the other is going slow.

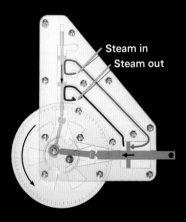

STEP 6 At this point the piston is coming to rest as it prepares to change direction, and the valve, traveling quickly in the downward direction, is just about to cut off the steam flow again.

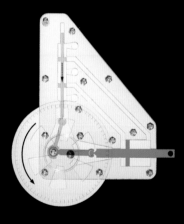

STEP 7 Here the steam is again cut off for a moment. The valve is traveling downward as fast as it can, and the piston is stopped for an instant. Note that while the piston and valve are starting and stopping, the flywheel just keeps turning at a steady speed. These wheels are made to be *very* heavy so they have enough inertia to carry the motion smoothly through the dead spots where there is no steam helping.

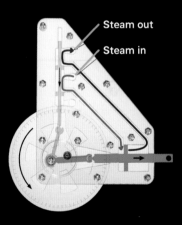

STEP 8 The valve has moved far enough to allow the steam flow once again into the lower pipe, which is where we started the cycle in Step 1. The piston is starting to accelerate toward the right, pushed by steam pressure from the left, thus completing a full cycle.

Harmonic Motion and Phase Angles

IF YOU WERE TO plot the up-and-down motion of the valve in the model we just looked at, you'd get a curve called a sine wave. The motion of an object following a sine wave is called *harmonic motion*. You will find this curve all over the place, representing all kinds of everyday movement. It's the motion of a vibrating string or of anything connected to the rotation of a wheel. It is also the curve followed by the voltage of alternating current, and by electric and magnetic fields associating with radio waves and light.

Harmonic motion can be described as the *smoothest possible* back-and-forth motion. The speed (velocity), the rate at which the speed is changing (acceleration), and the rate at which the acceleration is changing (jerk) all remain perfectly smooth throughout the cycle of the movement. In fact, all of these motions—the speed, the acceleration, the rate of change of acceleration, the rate of change of the rate of change of the acceleration, and so on forever—all follow sine waves, just shifted in time. (Mathematically this is described by the fact that the derivative of sine is cosine, which is just sine shifted by 90 degrees. No matter how many derivatives you take, it's sine waves all the way down.)

▽ **This graph shows the movement of the valve in the model on the prior page.**

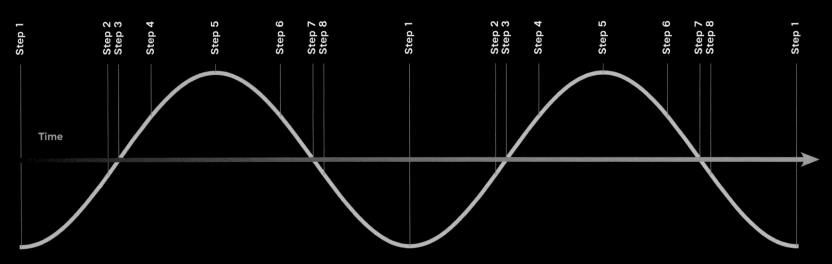

Step 1 Step 2 Step 3 Step 4 Step 5 Step 6 Step 7 Step 8 Step 1 Step 2 Step 3 Step 4 Step 5 Step 6 Step 7 Step 8 Step 1

Time

In Step 1 the valve is all the way at the bottom.

At Steps 2 and 3 the valve is about halfway to the top of its rotation and moving at its fastest rate of speed, represented by the steep slope of the line.

By the time we reach Step 5 the valve is at the top of its rotation, where it stops momentarily. This is shown by the top of the curve becoming perfectly flat.

At Step 6 the motion is reversed, as the valve begins moving back down.

The movements of the valve and the piston have a complementary relationship. This idea is illustrated most clearly if the movement of the piston is shown alongside that of the valve in the same graph. What the lines show is that when one is moving fast (steep slope), the other is slowing down, stopping, and turning around (flat curve). This is really the heart of what makes the design work. The valve needs to move fast to switch the steam when the piston needs to change direction, and at that moment the piston needs to stop so it can turn around.

Both the valve and the piston are following sine waves, but one is shifted by a quarter of a turn, or 90 degrees of rotation. These two waves together are commonly called sine and cosine waves, and we say that they are 90 degrees *out of phase* with each other. You can see in the model exactly where the 90-degree shift comes from. The movements are one-quarter cycle out of phase because the connecting rods rotated by 90 degrees—one-quarter of a circle—from each other. One of them responds to left/right movement of the crank, and the other responds to up/down movement.

It is a deep and fundamental fact about circles, and at the heart of the definition of the sine and cosine functions, that if you combine a sine wave in one direction with a cosine wave in the other direction, you get a circle. This is an incredibly general idea with applications all over the place. It is as fundamental to electric power and electric motors as it is to steam engines, as we'll see later on.

▽ Here is the same graph of the valve's movement from before (yellow), with the corresponding movement of the piston added (green).

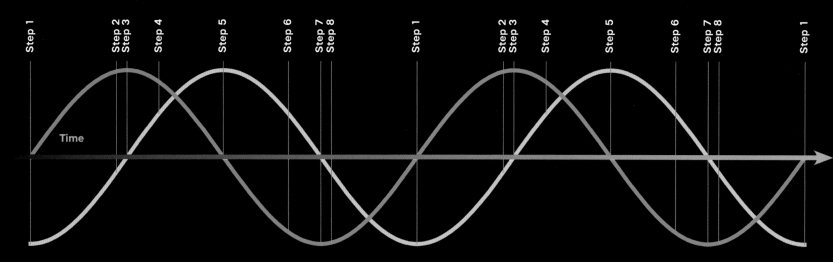

At Step 1 the piston is halfway through its movement, traveling toward the right.

At Steps 2 and 3 the piston is turning around at the far right, while the valve is moving as fast as it can through the center of its movement.

At Step 5 the piston is moving as fast as it can toward the left, illustrated by the steep descent of the line, while the valve is slowing and approaching its turnaround point.

Approaching a Real Steam Engine

THE ENGINE we just studied was designed specifically to make clear the 90-degree, quarter-turn relationship between the movement of valve and piston. This relationship is the same for all steam engines with this style of valve. But real steam engines (and every other model I know of) don't look anything like my abstracted model. That's because the 90-degree model we studied would make a *terrible* actual steam engine. For efficiency you want the steam valves to be as close as possible to the piston, preferably mounted right on the cylinder itself, letting steam directly into the cylinder with no pipes at all. Let's see how we can modify my model to achieve a more efficient and more realistic result.

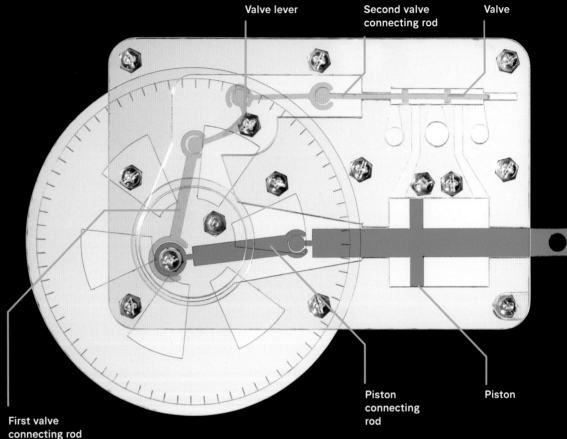

Valve lever

Second valve connecting rod

Valve

First valve connecting rod

Piston connecting rod

Piston

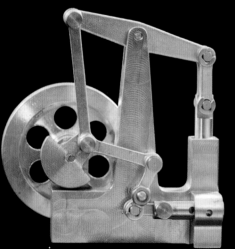

◁ This lovely little (less than 2 inches/5 cm tall) brass steam engine uses a very similar design, including the right-angle takeoff and motion-reducing lever. It's different in that it folds the motion of the piston rather than the motion of the valve, and it's slightly simplified in that the valve only sends steam to one side of the piston. So, like a person pedaling a bike, it only pushes, never pulls. That works but gives you only half as much power from the same size piston.

△ This model preserves the 90-degree relationship between the piston and valve connecting rods from the original model. It necessarily preserves the same timing of movement, but I've added an L-shaped lever that folds the motion over to bring the valve closer to the piston. The lever also allows the valve to move only half as far, making the whole model more compact.

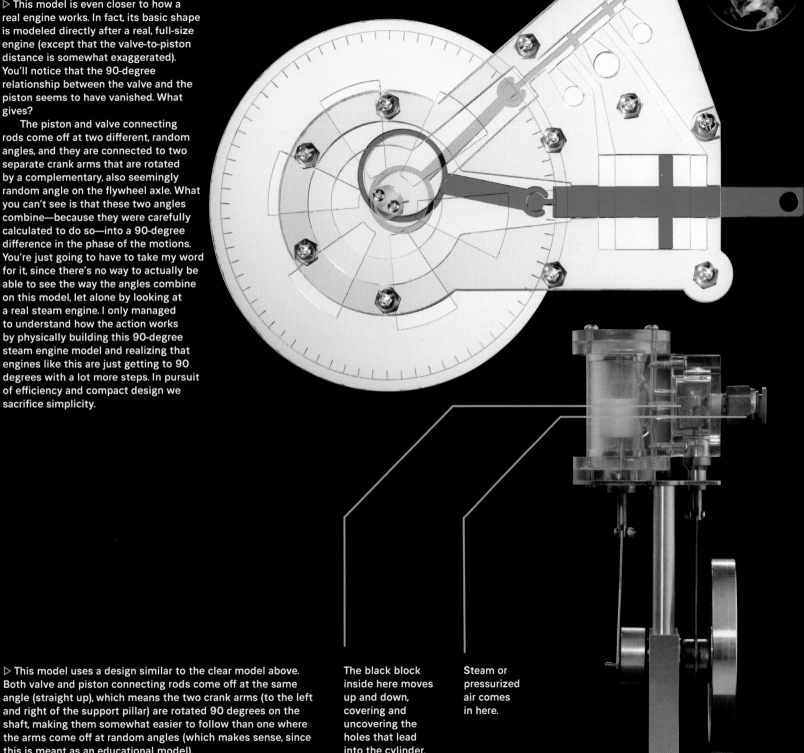

▷ This model is even closer to how a real engine works. In fact, its basic shape is modeled directly after a real, full-size engine (except that the valve-to-piston distance is somewhat exaggerated). You'll notice that the 90-degree relationship between the valve and the piston seems to have vanished. What gives?

The piston and valve connecting rods come off at two different, random angles, and they are connected to two separate crank arms that are rotated by a complementary, also seemingly random angle on the flywheel axle. What you can't see is that these two angles combine—because they were carefully calculated to do so—into a 90-degree difference in the phase of the motions. You're just going to have to take my word for it, since there's no way to actually be able to see the way the angles combine on this model, let alone by looking at a real steam engine. I only managed to understand how the action works by physically building this 90-degree steam engine model and realizing that engines like this are just getting to 90 degrees with a lot more steps. In pursuit of efficiency and compact design we sacrifice simplicity.

▷ This model uses a design similar to the clear model above. Both valve and piston connecting rods come off at the same angle (straight up), which means the two crank arms (to the left and right of the support pillar) are rotated 90 degrees on the shaft, making them somewhat easier to follow than one where the arms come off at random angles (which makes sense, since this is meant as an educational model).

The black block inside here moves up and down, covering and uncovering the holes that lead into the cylinder.

Steam or pressurized air comes in here.

More Than One Way to Drive a Steam Engine

THE ENGINES we just discussed all follow the same basic idea of two coordinated harmonic motions 90 degrees out of phase. But there are myriad other ways steam engines can be made to work. Over the centuries leading up to steam engines' final days of dominance in the mid-1900s, valve designs became ever more complicated but also more efficient, resulting in better, more powerful, more fuel-efficient engines.

We'll look at some of those "high-tech" steam engines in a bit, but first I wanted to see if I could make a variation that I had not seen anywhere else: a steam engine with only a single connection to the flywheel. At first I thought it must be impossible, because the valve needs to be moving when the piston is stopped. How can you make a mechanical linkage that moves when the thing driving it isn't moving? My solution was to amplify the motion *just before* the piston comes to rest.

It turns out this design was used in some steam locomotives, so it's not new, just new to me. But still, I'm pleased with my design, and it gives me a chance to put a bit more math into this book, which is never a bad thing. (Except with respect to sales. I read somewhere that for every mathematical formula included in a book, sales of that book will be cut in half. I can only assume this theory isn't more widely known because it is itself a formula, and thus doomed sales of the book it was in. So I'm only putting in graphs, not formulas.)

Let's step through the operating stages of my single-takeoff design.

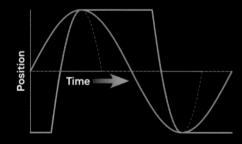

△ Using the same type of sine graph as before, we can visualize the motion of the valve as a sort of clipped, amplified sine wave. During most of the piston's movement (green line), the valve (orange line) isn't moving at all. But just before the top and bottom of the piston's curve, when the piston reaches the end of its motion, the valve jerks rapidly to one side or the other.

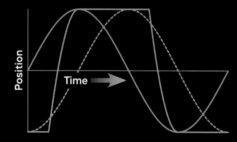

△ If we superimpose the cosine (shifted sine) wave from the right-angle models, we can see that the clipped curve is sort of an approximation of a cosine wave, but squared off.

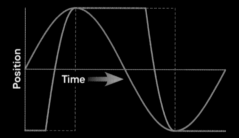

△ The squared-off cosine wave might seem like an inferior version of the original model, but if you think about it, what you *actually want* for maximum efficiency of the engine is for the valve to snap instantly from one side to the other, at the exact moment that the piston reaches the end of its travel. The optimal valve movement would in fact be a square wave, and the clipped movement we've created with the lever is a *better* approximation of this ideal than is the smooth sine wave. (The sine wave, on the other hand, has the virtue of being smooth and continuous, which makes the engine quiet and satisfying to watch, not all clickety-clacky.)

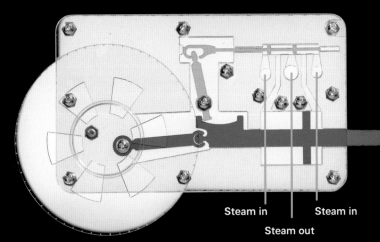

Steam in   Steam in

Steam out

STEP 1 At this stage the lever arm has pushed the valve all the way to the left and the piston is stopped at the limit of its motion, pushed all the way to the right. Steam flows from the right steam inlet to begin to push the piston toward the left.

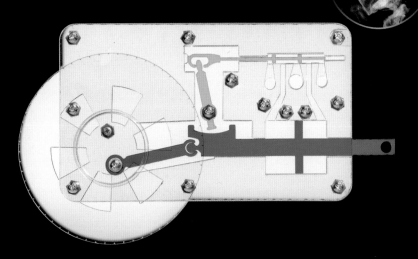

STEP 2 As the piston is forced left by the steam, the lever remains stationary, held in place by friction as steam continues to flow into the right port and out the center port.

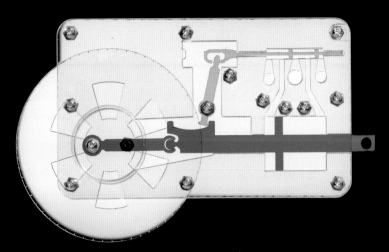

STEP 3 When the piston arrives at its maximum point of left-bound motion, it jerks the lever over, which in turn moves the valve to the far right and now forces the steam to flow from the left port, pushing the piston to the right.

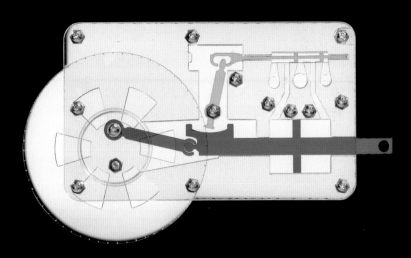

STEP 4 As the piston moves back to the right, the lever once again remains stationary, held by friction, until the piston arrives back on the right and jerks the lever to the opposite direction, which puts us back at Step 1.

The Simplest Steam Engine

NOW THAT I'VE bored you with my single-connection design, here's an example of a single-takeoff design that is widely used, even today. The simplest steam engines, both model and real-world, connect the piston to the crank arm with a fixed, rigid rod. When the crank arm moves back and forth, the whole piston and the cylinder around it swing back and forth so the piston stays pointing toward the end of the crank.

This would not be realistic in a large engine where the piston may weigh several tons, and the cylinder around it many more tons. But in models and in small, low-power real engines, it greatly simplifies the whole design. The cylinder itself becomes the valve, and the side-to-side movement of the cylinder opens and closes steam channels as needed.

Mathematically speaking, my single-takeoff design was complicated by the fact that I was not only making a single connection, I was also using only a single direction of movement, just the in-out movement of the piston along a fixed axis. These engines also have a single connection, but they use *two* directions of movement. One direction of movement is used to push the piston in and out, while the other direction of movement is used to tilt the cylinder back and forth.

A disadvantage of this design is that it can only be used for "single-acting" cylinders. That means the steam pushes only in one direction, and the piston has to rely on the attached flywheel to provide enough inertia to coast back in the other direction, pushing out the leftover steam. For this reason, it's common to see this design used in two-cylinder engines, with the two separate cylinders pushing alternately to keep constant pressure on the flywheel.

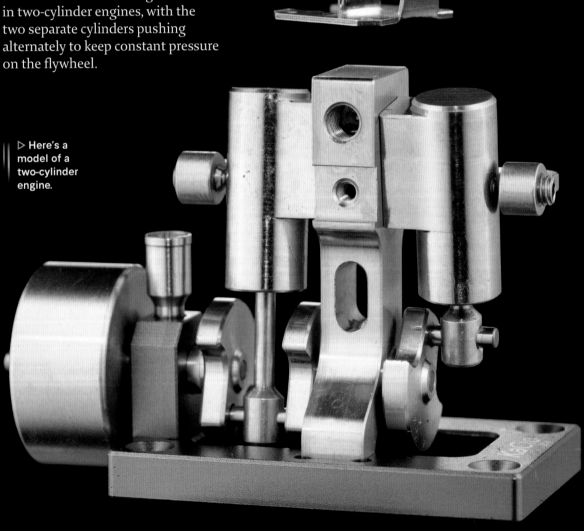

▷ Here's an alternate arrangement of a two-cylinder, sliding-valve steam engine. In this example both cylinders push on the same crank arm, but they are at a 90-degree angle from each other, so they push at different times.

▷ Here's a model of a two-cylinder engine.

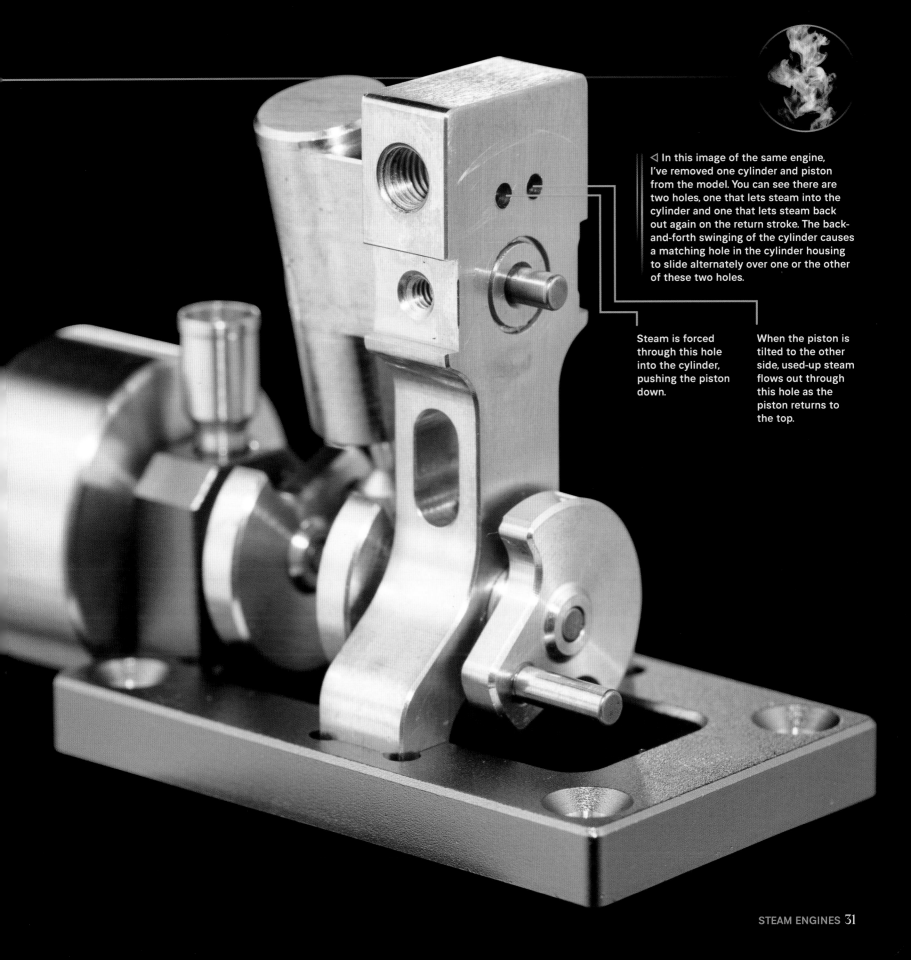

◁ In this image of the same engine, I've removed one cylinder and piston from the model. You can see there are two holes, one that lets steam into the cylinder and one that lets steam back out again on the return stroke. The back-and-forth swinging of the cylinder causes a matching hole in the cylinder housing to slide alternately over one or the other of these two holes.

Steam is forced through this hole into the cylinder, pushing the piston down.

When the piston is tilted to the other side, used-up steam flows out through this hole as the piston returns to the top.

The Best of the Old School

YOU CAN GO pretty far down the rabbit hole trying to make the most efficient possible steam engine valve—and people did. The Corliss design, patented in 1849, was a major step forward in efficiency. Instead of one valve it has four, including a separate inlet and outlet valve—of different types—for each side of the piston. This far more complicated (and thus more expensive) design is worth the cost because it allows the engine to deliver more power with less fuel and water consumption.

To understand how this works, consider the moment in our earlier models at which the valve switches steam from one side of the cylinder to the other. When the valve closes, all of the high-pressure steam the cylinder had been receiving is immediately vented to the outside,

so a whole cylinder's worth of steam is just dumped, wasting a *lot* of steam and the fuel used to heat it up. Although the governor reduces the pressure to slow down the engine, this does nothing to change the fact that whatever pressure you have left is wasted at the end of each half-cycle.

In the Corliss design, steam is fed constantly at full pressure into the cylinder, and the speed of the engine is instead regulated by adjusting how long the inlet valves stay open. Under low load, the valves snap open for just a fraction of a second right at the start of the stroke, after which the steam expands in the cylinder on its own. By the end of the stroke the pressure inside is much lower, so there is less waste when the exhaust port opens.

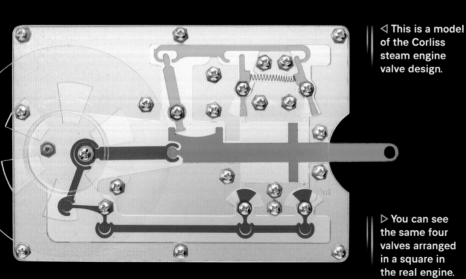

◁ This is a model of the Corliss steam engine valve design.

▷ You can see the same four valves arranged in a square in the real engine.

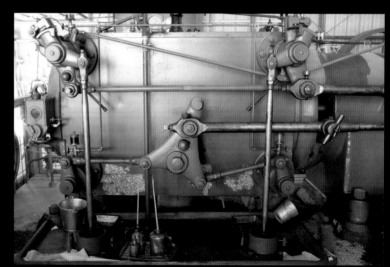

◁ This immense engine, the Cooper Corliss, is located at the Silver Creek Museum in Freeport, Illinois.

They miscalculated the height of the roof and had to cut a notch out of the beam.

The Corliss engine was not the end of the line for steam engine improvements. New and improved steam engines continued to be built even after internal combustion engines had become well established, just like cars powered by gasoline engines are still on the road even though electric vehicles are well on their way to displacing them.

Steam-powered cars had real advantages over those run on internal combustion engines. For example, they had no transmission, because they can generate high torque (turning force) down to zero rpm (revolutions per minute). They were also quieter and, for a while at least, faster. But in the long run, steam engines just could not keep up and internal combustion engines won the century, as we will learn in the next chapter.

◁ This Doble E-20 Steam Car, built in 1925, recycles its steam so you don't have to add water. It's got electric motors (to pump water, fuel, and oil) and an electronic control system that regulates the fuel flow to the boiler. It even runs on gasoline!

Engines That Spin
Without Pistons

PISTON-DRIVEN reciprocating steam engines that actually run on steam, like the ones we've just been looking at, are now found only in museums, at antique farming shows, and in the backyards, workshops, and showrooms of hobbyist collectors. But that doesn't mean steam engines are dead—far from it. Some of the most powerful steam engines that have ever existed—which are some of the most powerful engines of any kind in the world today—are steam turbine engines operating inside power plants.

Nearly all conventional, utility-scale power plants are based fundamentally on steam turbine engines. They get their steam the old-fashioned way, by boiling water; the only difference is where the heat comes from. A coal-fired power plant burns coal. Oil-fired plants burn oil. Nuclear power plants generate heat from nuclear reactions. Each of these sources heats water to a boil, creating steam at very high pressure. This steam is then directed through a turbine engine, which spins an electric generator.

Turbines capture energy from steam the same way a windmill captures energy from moving air (which is why windmills are often called wind turbines when they're being used to generate electricity). A windmill lets most of the air go past it, because who cares—wind is free! Steam turbines, on the other hand, are designed to capture as large a fraction as possible of the energy in the steam flowing through them. The energy you can get from a flowing liquid or gas is the energy released by a decrease in the pressure of the material flowing past the turbine blades. This decrease might be tiny, as in the case of a wind turbine out in the open air; or it might be huge, as in the case of a turbine in a power plant where ultra-high-pressure steam enters on one side and exits at nearly atmospheric pressure on the other side (the lower the exit pressure, the more efficient the engine). Efficient turbine engines have multiple stages, sometimes dozens of them, that the steam has to flow through before it can escape. Each stage is slightly different, optimized for the pressure, volume, and speed of the steam that will be passing through it. The steam slowly cools and depressurizes, releasing a bit more energy at each stage.

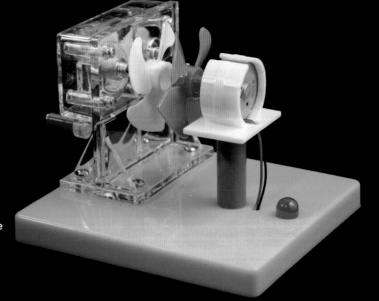

▷ A turbine engine is basically a fan working backward. Instead of using a motor to turn a fan to blow air, a turbine uses blowing air to turn a fan, which turns an electric generator. This model has one of each pointing at each other: when you turn the crank on the back, the white fan spins, blowing air at the red fan, which acts like a wind turbine and starts turning as well.

▷ While a fan out in the open is technically a turbine, in many cases it's more efficient to put the fan blades into a housing that contains and concentrates the flow of whatever is driving the turbine. The shape, number, and spacing of the fan blades depend on whether the turbine is meant to run on air, steam, burning fuel, or water, and how much flow and pressure is expected.

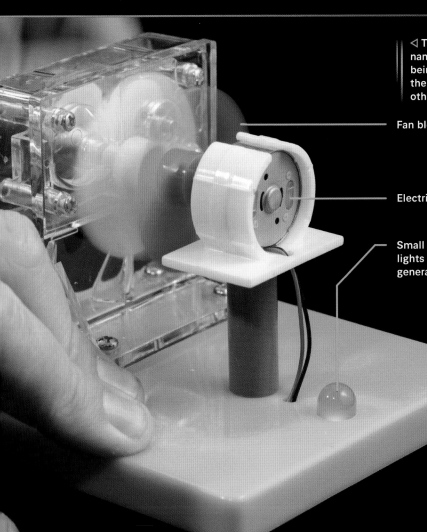

◁ Turbine—a fancy name for a fan—being blown by the air from the other fan

Fan blowing air

Electric generator

Small LED that lights up when the generator is turning

△ Giant windmills, known more formally as wind turbines, are just overgrown versions of the toy on the left. They use naturally occurring wind—a form of solar energy—to turn a big fan. Windmills like these near my farm in rural Illinois spend a lot of time out on the lonely prairie in the unending wind that has been known to drive people insane. We don't know much about how windmills cope with life, but at least we know what kind of music they prefer. They are huge metal fans.

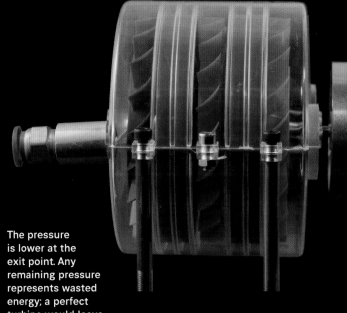

The pressure is lower at the exit point. Any remaining pressure represents wasted energy; a perfect turbine would leave the exhaust lazily falling out.

High-pressure gas or liquid enters here.

△ Multiple stages capture more energy from the pressurized steam before it's finally released on the far end.

STEAM TURBINES for nuclear power plants are absolutely huge, and fantastically expensive. The solid steel shaft down the middle of this one is around 18 inches (46 cm) in diameter, to give you an idea of the scale of this thing. It's almost as long as a football (or soccer) field. A single such turbine can generate over 1,700 megawatts (1.7 gigawatts) of power, enough to supply about 1.7 million homes on average.

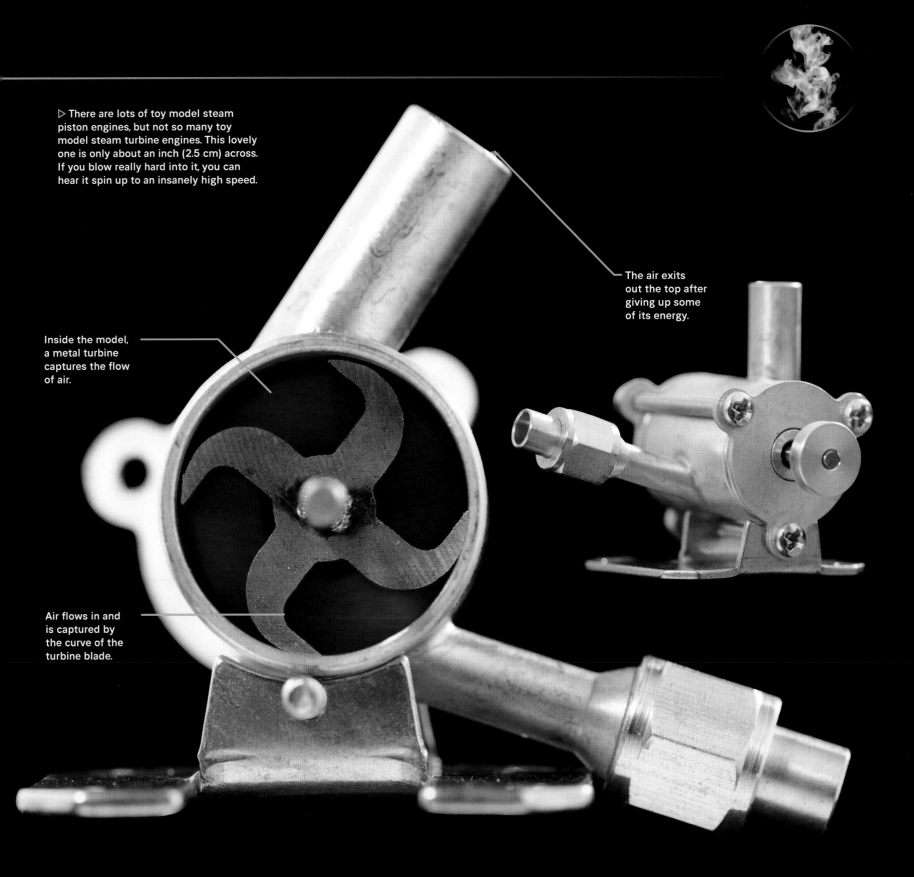

▷ There are lots of toy model steam piston engines, but not so many toy model steam turbine engines. This lovely one is only about an inch (2.5 cm) across. If you blow really hard into it, you can hear it spin up to an insanely high speed.

The air exits out the top after giving up some of its energy.

Inside the model, a metal turbine captures the flow of air.

Air flows in and is captured by the curve of the turbine blade.

Engines That Work by Throwing Something Out the Back

▽ A modern jet engine gets most of its thrust from the air blown *around* the engine, not from the hot gas shooting out the back. (Jets that don't have this bypass fan do exist, but they are very wasteful of fuel.)

THE AIR HOGS toy engines that we looked at earlier are examples of engines that use a high-pressure gas to spin a propeller that pushes an airplane through the air. But why not just allow the high-pressure gas to shoot out the back in order to propel the plane? There are engines that work this way, but they have some serious disadvantages.

Imagine you are floating on a lake in a motorless boat with no wind. How do you get the boat to move? You could try blowing really hard off the back, and you might get it to move a *tiny bit*, but you're going to collapse before you get anywhere. Believe it or not, this is exactly how a rocket engine works. In order to launch, it throws hot gas very hard out the bottom of its cylinder and reaches space by the equal and opposite reaction force. But rocket engines are *incredibly inefficient*. The amount of fuel needed to lift a rocket a few miles into the air would fly you around the world in an airplane.

Now imagine that your boat is full of younger siblings, each weighing 45 pounds, and you throw them as hard as you can off the back of the boat. This will actually get you going. Each kid you throw creates an equal and opposite reaction,

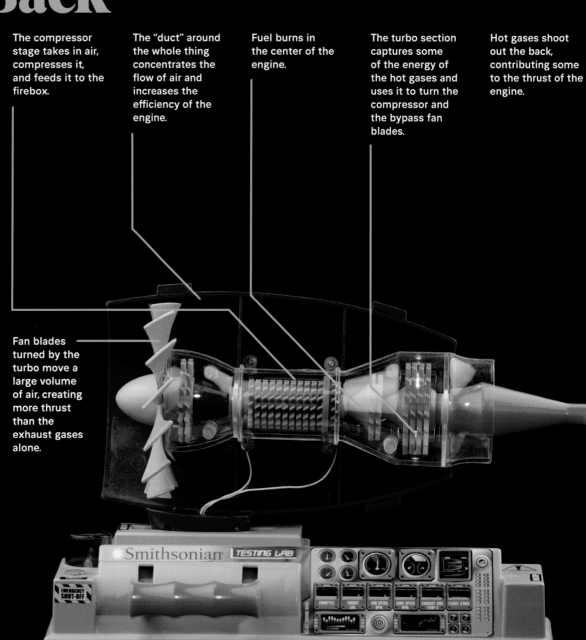

The compressor stage takes in air, compresses it, and feeds it to the firebox.

The "duct" around the whole thing concentrates the flow of air and increases the efficiency of the engine.

Fuel burns in the center of the engine.

The turbo section captures some of the energy of the hot gases and uses it to turn the compressor and the bypass fan blades.

Hot gases shoot out the back, contributing some to the thrust of the engine.

Fan blades turned by the turbo move a large volume of air, creating more thrust than the exhaust gases alone.

▷ All the great dogfighters of WWI and WWII used piston engines, but modern propeller planes are often powered by turbine engines instead. These are known as turboprop planes. They are still in common use, particularly for smaller airplanes used on short runs. Their engines are basically identical to jet plane engines, but without a duct around the propellers. These planes are noisy and can't fly as fast as jets, but they can actually be more fuel efficient.

▽ Unlike the soda bottle rocket, real rockets are examples of engines that work by blowing just plain gas out the back. They get off the ground only because they throw out a *lot* of *very fast*-moving gas. Although it's hard to compare the power of any rocket engine with other kinds of engines, no matter how you calculate it the power of the Saturn V moon rocket, shown here, blows away all other engines ever made, before or since. This machine is more powerful than the biggest gas turbine engine, the biggest diesel engine, and the biggest electric motor, all put together and multiplied several times.

pushing the boat forward. You could realistically get maybe 8 to 10 feet (3 m) per kid. But when you run out of siblings, you're stuck—and soon you're going to have several angry kids climbing back into your boat.

Finally, imagine you have a paddle or even two oars, and can use those to push off against the water around you. That's going to get you moving with *far* less effort than either blowing air or throwing kids! An airplane propeller is just a paddle adapted for the air. Explained in terms of mass and momentum, the propeller is using the plentiful supply of mass in the air all around it to impart momentum to the plane. The best part is that the plane doesn't have to carry any

of this mass on board. Just like you don't have to carry a bunch of kids on board with you to create momentum for your boat by using paddles. You can keep paddling, and moving forward, as long as you have energy to drive the paddle and lake to float on.

You might think that jet engines work the same way rocket engines do, by throwing hot gas out the back. To some extent they do, but jet planes are a lot more similar to propeller planes than they appear. Jet engines are a combination of a turbine engine, a compressor, and a "ducted fan," which is basically a propeller in a tube.

Jet fuel is burned in a firebox located in the middle of the engine.

This creates hot, pressurized gas, which drives a turbine on the back side of the engine. Connected to the same shaft as the turbine is a compressor located on the front side of the engine, which sucks and compresses incoming air, allowing more fuel to be burned, and thus generating more power.

Also connected to the same shaft is a much larger set of propeller blades at the very front of the engine, which blows air out the back, completely bypassing the engine in the middle. The main difference between a jet plane and a turboprop (propeller) plane is that in the jet plane these large blades are inside a tube, while in the propeller plane they are on the outside.

▷ This is a rare example of an engine that works by the "throwing kids off the back of a boat" principle. To make it work, you need to fill half the bottle with water, then add compressed air with a bicycle pump. If you use either too much or too little water, it won't go as high. The compressed air supplies the energy when it expands, and the water supplies the reaction mass to impart momentum to the rocket. In the boat analogy, the compressed air is you, and the water is the kids you're throwing off the back. The range of this engine is severely limited because it has to carry all its reaction mass with it, rather than pushing off the air around it. The only real advantage of this type of engine is that it—like a real rocket—would work equally well on the moon or in outer space, where there is no air to push against.

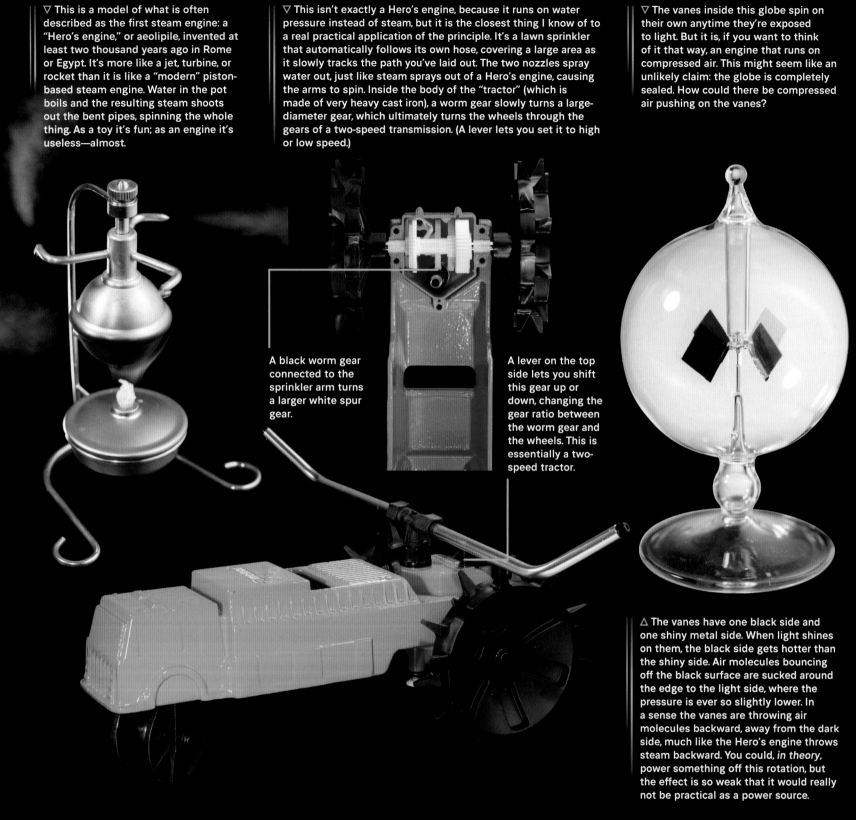

▽ This is a model of what is often described as the first steam engine: a "Hero's engine," or aeolipile, invented at least two thousand years ago in Rome or Egypt. It's more like a jet, turbine, or rocket than it is like a "modern" piston-based steam engine. Water in the pot boils and the resulting steam shoots out the bent pipes, spinning the whole thing. As a toy it's fun; as an engine it's useless—almost.

▽ This isn't exactly a Hero's engine, because it runs on water pressure instead of steam, but it is the closest thing I know of to a real practical application of the principle. It's a lawn sprinkler that automatically follows its own hose, covering a large area as it slowly tracks the path you've laid out. The two nozzles spray water out, just like steam sprays out of a Hero's engine, causing the arms to spin. Inside the body of the "tractor" (which is made of very heavy cast iron), a worm gear slowly turns a large-diameter gear, which ultimately turns the wheels through the gears of a two-speed transmission. (A lever lets you set it to high or low speed.)

▽ The vanes inside this globe spin on their own anytime they're exposed to light. But it is, if you want to think of it that way, an engine that runs on compressed air. This might seem like an unlikely claim: the globe is completely sealed. How could there be compressed air pushing on the vanes?

A black worm gear connected to the sprinkler arm turns a larger white spur gear.

A lever on the top side lets you shift this gear up or down, changing the gear ratio between the worm gear and the wheels. This is essentially a two-speed tractor.

△ The vanes have one black side and one shiny metal side. When light shines on them, the black side gets hotter than the shiny side. Air molecules bouncing off the black surface are sucked around the edge to the light side, where the pressure is ever so slightly lower. In a sense the vanes are throwing air molecules backward, away from the dark side, much like the Hero's engine throws steam backward. You could, *in theory*, power something off this rotation, but the effect is so weak that it would really not be practical as a power source.

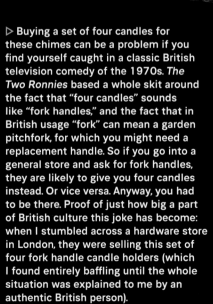

◁ This lovely set of Christmas chimes is a gas turbine engine that runs on the power of four candles. Hot air rises (because it is less dense than cold air), and the upward flow from the candle flames is intercepted by the fan blades near the top. This chime is not even close to efficient, but what it lacks in efficiency it more than makes up for in peaceful beauty that reminds us of a simpler time.

▷ Buying a set of four candles for these chimes can be a problem if you find yourself caught in a classic British television comedy of the 1970s. *The Two Ronnies* based a whole skit around the fact that "four candles" sounds like "fork handles," and the fact that in British usage "fork" can mean a garden pitchfork, for which you might need a replacement handle. So if you go into a general store and ask for fork handles, they are likely to give you four candles instead. Or vice versa. Anyway, you had to be there. Proof of just how big a part of British culture this joke has become: when I stumbled across a hardware store in London, they were selling this set of four fork handle candle holders (which I found entirely baffling until the whole situation was explained to me by an authentic British person).

Modern Steam Engines
That Don't Run on Steam

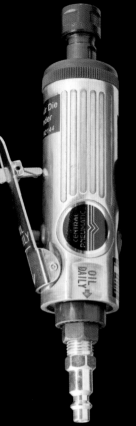

STEAM-DRIVEN piston engines are museum pieces, and steam turbine engines are found only in huge power plants. But engines that run on compressed air are all over the place. If you extend the definition to include engines that run on compressed liquids (such as hydraulic fluid), there are even more to be found on construction sites and farms, and in factories and workshops.

Handheld air tools nearly all use "vane" motors. There is no piston or flywheel, just a set of paddles that is pushed around by the flowing air. This type of motor could run on steam, but it needs a lot more compressed gas than a piston engine to deliver the same amount of output power, so the design was never used with actual steam. A vane motor comes in handy when light weight and mechanical simplicity are more important than efficiency.

Vane motors can be made *very* cheap. The air die grinder on this page was on sale for $9, which is less than I paid for the sandwich I had for lunch the day I got it! (To be fair, it was a pretty good sandwich.) That's because they are designed so almost no precision machining is necessary. The ball bearings are the only part of the tool that has to be made to high precision, and those are commodity components manufactured in vast quantities, so they too can be made cheaply using highly specialized tooling. The rest of the tool is made to tolerances easily achieved by beat-up old factory equipment.

△ An air die grinder spins very fast and is typically used with a small grindstone or diamond burr to smooth edges or remove unwanted mold marks from metal castings.

▷ The vane motor inside the air die grinder has four dark steel vanes that spin inside the hollow slotted tube. The vanes have to form a tight seal to the inside of the tube, or air will leak around them. This could be done by manufacturing both the tube and the vanes to very high precision, but it's much cheaper to make the vanes fit loosely in slots, so they can move in and out easily (the one in the bottom slot has fallen out, and the one in the top slot is all the way down in the slot, so we see only the one facing toward us). When the motor is spinning fast, centrifugal force presses the vanes out and keeps them tight against the inside of the tube.

The collet is like a circular clamp that holds the shaft of a small grindstone, drill bit, or diamond wheel.

The vanes fit in these slots in the rotor.

This steel vane belongs in the slot that's facing down, but it has fallen out because the vanes fit very loosely.

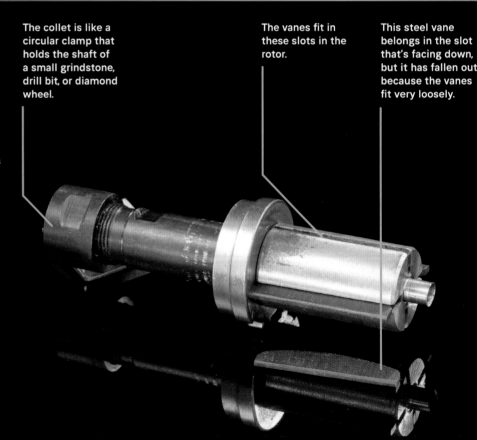

This is an air impact wrench, the tool of choice for taking wheels off cars or putting them back on. If you've ever spent time in a car repair shop, you have heard one of these in action. When taking a nut off, it starts out sounding like someone is hammering really fast, then transitions into a sort of modified dentist's drill sound. The job of these tools is to turn a nut or bolt, but instead of just applying steady pressure (as you would with a hand wrench), they hammer on the nut several times a second until the nut comes around to seeing the wrench's point of view. Using compressed air allows a small, lightweight tool to hit the nuts and bolts very hard (using a sort of internal rotary hammer mechanism). Once the nut is loose, the tool switches to spinning fast to move it along.

There's a reason impact wrenches sound a bit like dental drills: Both run on compressed air. The dental drill is much smaller, runs at a much higher speed, and doesn't have an integrated hammer mechanism (**YOU HOPE**).

These are both random orbit sanders. The one on the right runs on compressed air, and the one on the left is electric. One fascinating difference you notice after using each is that the electric motor gets warmer the longer you use it, but the air motor gets *colder*. It can actually get so cold it's uncomfortable to hold. That's because air, like most gases, cools down when it expands. The mechanical energy delivered by an air motor comes from the expansion of air happening inside the motor, so that's what gets cold.

Air exits through these slots after it's pushed the vanes about halfway around the circle.

High-pressure air enters through this port, on the opposite side from the exit slots in the hollow tube.

This clever little groove directs a small amount of pressurized air to the space behind the vanes in the rotor, helping press them outward. This gets the tool started before centrifugal force can keep the vanes pressed outward.

Air cools when it expands, but that means it also heats up when it's compressed. Notice the cooling fins and fan designed to keep this compressor head cool.

Vacuum Motors

MOTORS THAT RUN on compressed air are common, but there is a less common type that runs on the opposite: a vacuum. A common example in earlier decades was the motor that ran the windshield wipers on most cars (today most are electric). But by far the most beautiful example of a vacuum motor is the kind found in classic player pianos.

These beautiful instruments, popular from the late 1800s to the mid-1900s, operate entirely on vacuum. Foot pedals operate bellows that create negative pressure in a chamber in the base of the piano. From there hoses lead up to all sorts of valves, bellows, and controls that turn the music rolls, read the punched holes, and press the keys.

Because we're interested in motors here, our focus is on the astonishing (but very common in its day) five-bellows vacuum motor that turns the music roll. It looks like some kind of weird insect with five legs, all moving up and down in a wave from one end to the other. These legs move sliding valves, which alternately connect the bellows first to the open air, then to the vacuum chamber below. When the bellows are connected to the air, they relax and open up. When they're connected to the vacuum, they are sucked together, driving the crankshaft just like the pistons in a steam engine. Why five bellows? To make the movement as smooth and steady as possible, with several always pulling at the same time, and handing off gently to the next one in line. Why does the vacuum motor in *my* player piano *not* turn smoothly? Because it hates me, that's why. I've tried everything.

▷ Windshield wipers need to go back and forth in an arc as they wipe away rain. This one could work with air from a compressor pushing the vane first on one side, then the other, but it doesn't. Instead, it uses "engine vacuum"—suction created as a side effect of the movement of the pistons in a gasoline or diesel engine. This vacuum is applied alternately to one side, then the other, sucking the vane back and forth. (Technically it's the pressure of the atmosphere on the other side that pushes the vane, but the point is that this device is operating on less-than-atmospheric pressure, instead of more-than-atmospheric pressure like all steam and compressed air engines.)

This vane fits tightly in the half-circle case around it. The wiper blade is attached directly to the shaft coming out of the motor, so when the vane swings back and forth, the wiper sweeps across the windshield.

When the vane reaches one end, it flips this toggle, directing the vacuum to the other side, which reverses the direction of the motor.

▷ From the front we see the sliding valves, which cover and uncover port holes.

When the hole is exposed, the bellows beneath can open up as surrounding air flows in.

When the port is covered, vacuum sucks the bellows closed.

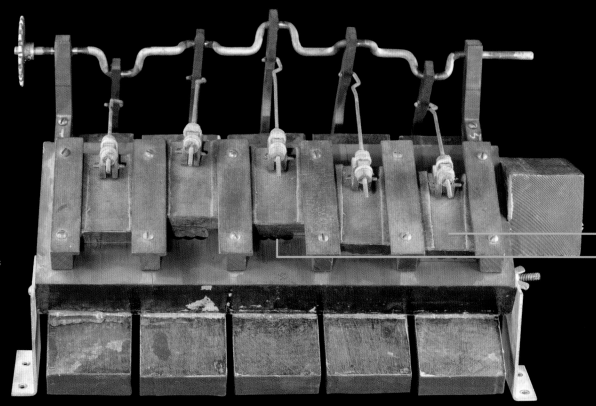

Closed valve

Open valve

▷ From the back we see the bellows, and the connecting rods that join them to the crankshaft. This is like a five-cylinder steam engine, except instead of steam pressure pushing on pistons, there is vacuum suction compressing bellows.

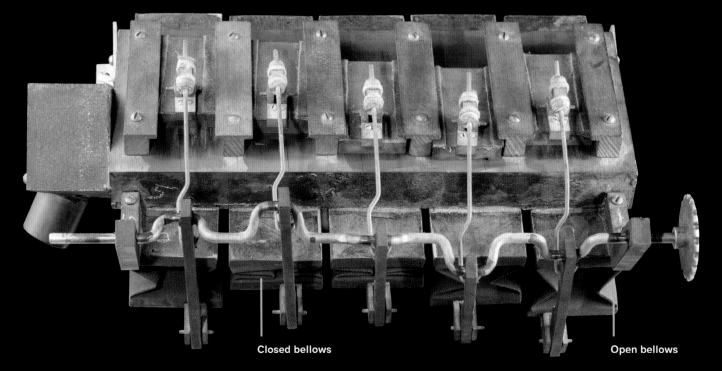

Closed bellows

Open bellows

Hydraulic Motors

MOTORS THAT RUN with hydraulic fluid instead of air are also quite common. Their main advantage is that they can deliver an *enormous* amount of torque (turning force) at any speed, with very precise control. Hydraulic motors can deliver so much force because the fluid coming into them is at such a high pressure (several thousand pounds per square inch). Oil flows in one side under high pressure and out the other at much lower pressure.

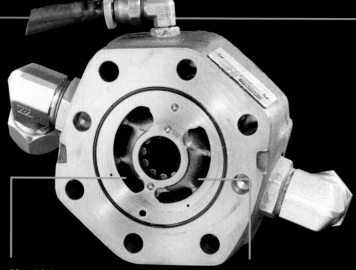

Very high-pressure oil enters here.

Much lower-pressure oil exits here.

This is the case drain, which leads any oil that may leak internally back to the reservoir tank to be pumped back through the loop.

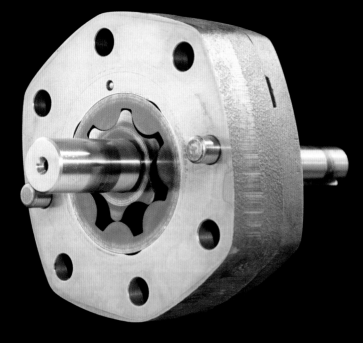

△ This is a "gerotor" type motor, which is quite similar in some ways to a vane motor, except this one *is* manufactured to very high precision (and costs something like fifty times as much). Carefully shaped ports on one side feed high-pressure oil into the space between the teeth in the gears. The inner gear, which has one less tooth than the outer gear, is forced to turn by the pressure. As it rotates, the gear delivers the oil one tooth-load at a time to the low-pressure outlet port on the other side. The shape of the ports is carefully designed so that as the teeth pass by, a steady pressure and flow rate are maintained, minimizing vibration.

△ Notice how incredibly *thick* all the parts are. This motor, despite being only about 8 inches (20 cm) across, is difficult to lift, being made almost entirely of solid steel. That's because the hydraulic fluid coming in is at very high pressure, several thousand pounds per square inch (20 MPa [megapascals], or about 200 times atmospheric pressure). The total force on the internal parts and housing is *huge*.

Where Air and Hydraulic Motors Are King—and Remind Us of Steam Engines

NOT FAR FROM where I live there is a community of Amish who have decided that paying an outside company to deliver electricity to their homes and workshops is, on the whole, a bad idea. But many of them make their living manufacturing furniture and cabinetry with large industrial woodworking machinery that ordinarily uses electric motors. Their solution to this dilemma is to replace all the electric motors on their machinery with hydraulic motors and run them all off a large central hydraulic pump powered by a diesel engine.

This makes a lot of sense: Hydraulic motors run without

generating heat and deliver smooth torque even better than electric motors. The main reason non-Amish shops don't use hydraulic motors is because it's simpler and more convenient to run electrical wires than hydraulic lines, and it's nice not to have a big diesel engine to maintain and listen to. But in terms of the machinery itself, the Amish solution is generally superior.

Let's take a tour of the shop at Four Acre Wood Products in Arthur, Illinois (which, incidentally, supplied much of the furniture and all the doors and flooring of the house I built about twenty-five years ago).

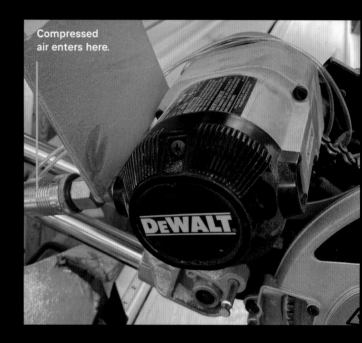

Compressed air enters here.

▷ At first glance this looks like a perfectly ordinary DeWalt brand sliding compound miter saw. I have this exact model myself! But look closely and there's something off about what's attached to the back of the motor housing.

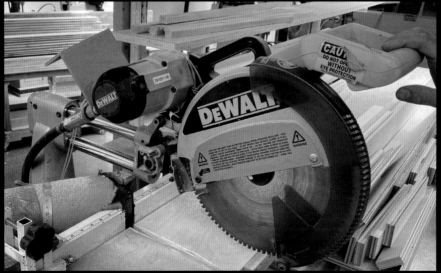

△ The electric motor that would normally be inside this plastic housing has been replaced with a compressed air motor that fits perfectly in the same space. What's attached to the side is an air line, not an electric cord. As a result, this saw sounds like an overgrown dentist's drill instead of like an angry vacuum cleaner.

△ This is a *really nice* panel saw. Seriously nice. I want this saw. But look closely and you'll see that I'd have trouble using it in my shop.

▷ Again, we see that the electric motor has been replaced, this time with a high-pressure hydraulic motor. Some manufacturers will sell this type of equipment with no motor at a discount, but if not, the Amish just remove the electric motor and, I suppose, resell it to the English (that's the non-Amish who live around them).

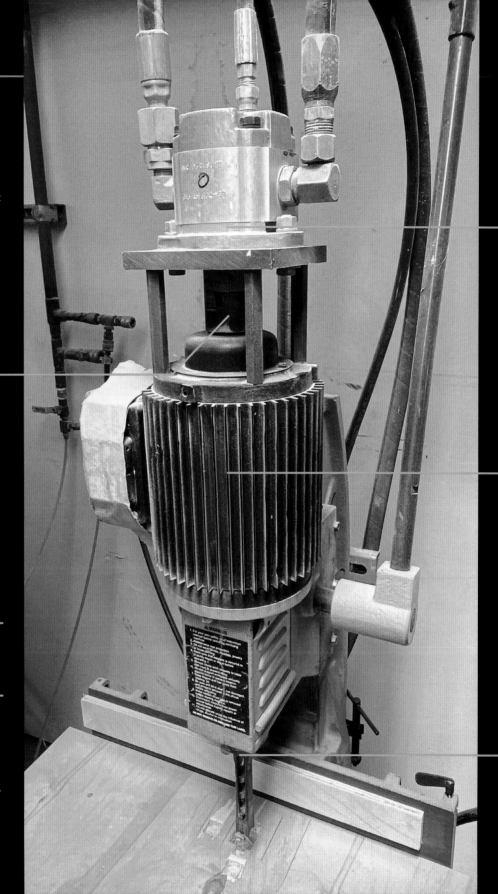

▷ Sometimes it's easier just to leave the electric motor in place instead of removing and replacing it, as is the case with this Frankensteined drill. The top end of the electric motor, which would normally drive an enclosed cooling fan, has been removed and replaced with a hydraulic motor, which drives the drill bit *through* the electric motor, which is just along for the ride.

This hydraulic motor drives the machine.

The coupling between the two motors is called a Lovejoy coupling, which consists of two Lovejoy hubs and a Lovejoy spider. It allows for a small amount of misalignment and provides a bit of shock-damping. Of all the motor coupling designs, the Lovejoy coupling has by far the best name.

This electric motor does nothing useful but its ball bearings probably help keep the shaft straight.

Hydraulic fluid enters here at around 3,000 pounds per square inch (20 MPa).

This is a mortising machine, which drills square holes using a round drill bit spinning inside a square chisel.

This smaller hose is the case drain, which collects and returns any fluid that leaks internally.

Hydraulic fluid leaves through this hose at much lower pressure.

▽ This table saw is of course hydraulically driven, but it's interesting for another reason.

◁ If your finger so much as brushes the spinning blade, this block of aluminum is slammed into the blade, stopping it in less than 5 milliseconds. The aluminum block and the blade are destroyed, but not your finger. Which is a fair trade.

△ The saw is fitted with the latest in high-technology safety features: an auto-stop. It is of course an electronic device. The Amish do not reject electricity unconditionally; they just restrict its use to places where they feel there is no better alternative—like, for example, not chopping your finger off. It has a sort of "black box" that records exactly the signal that caused it to trigger. Users are asked to return the triggered device to the company, along with a description of what happened. They have been using the black box recordings to steadily improve the ability of the circuit to distinguish between fingers and, for example, damp wood, nails, or other things that are less valuable than a blade and auto-stop module, and should thus not trigger it.

△ Machines get a lot more complicated than simple saws and drills. This is a multi-surface shaper. It has two shaper heads (like really big router bits), each of which can be moved in and out or up and down with powered actuators. Similarly the rollers on top lower under power to press down and drive boards through the machine. The control panel on the right has switches that control the various motions of the machine.

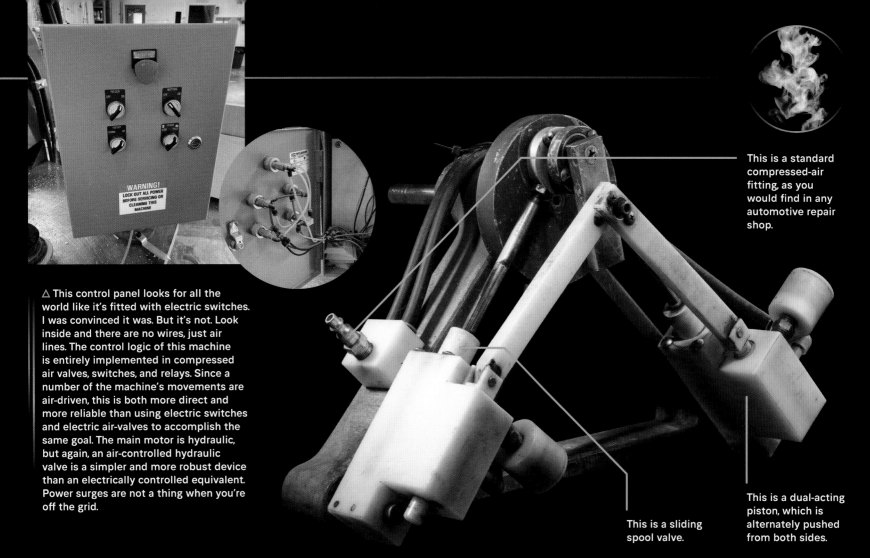

This is a standard compressed-air fitting, as you would find in any automotive repair shop.

△ This control panel looks for all the world like it's fitted with electric switches. I was convinced it was. But it's not. Look inside and there are no wires, just air lines. The control logic of this machine is entirely implemented in compressed air valves, switches, and relays. Since a number of the machine's movements are air-driven, this is both more direct and more reliable than using electric switches and electric air-valves to accomplish the same goal. The main motor is hydraulic, but again, an air-controlled hydraulic valve is a simpler and more robust device than an electrically controlled equivalent. Power surges are not a thing when you're off the grid.

This is a sliding spool valve.

This is a dual-acting piston, which is alternately pushed from both sides.

▽ Cooling fans run on air-driven piston engines. This one in the furniture shop's showroom has two air lines, so the exhaust air is directed out of the showroom. (Tools in the shop don't need this drain line, which both makes the motor quieter and prevents any lubricating oil from getting into the nice clean showroom.)

△ This is a two-cylinder version of the fan motor. It's a lot like the tiny model we saw on page 31, but many times bigger (about 10 inches tall). It's built in every way like a steam engine, but it's an air motor in active use, made new in 2020. It could almost certainly run on steam, but it's run on compressed air just because that's more convenient for the situation it's in.

△ The same piston engines run pedestal fans in the shop.

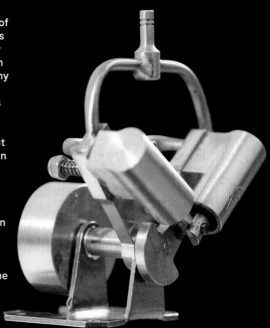

△ So where does all this hydraulic (liquid pressure) and pneumatic (air pressure) power come from? This is the engine room where a powerful diesel engine supplies all the power for the shop.

See the next chapter to learn all about internal combustion engines like this!

△ The alternative to piston engines for air power is a vane motor like this, also commonly used by the Amish. It's mechanically simpler and more reliable but uses a lot more compressed air to do the same work to turn a fan. The piston engine on the previous page is just about the most efficient way to turn a small amount of high-pressure air into a much, much larger amount of slow-moving, low-pressure air. That's why the piston design is used both in the showroom and in the Air Hogs toy.

▽ Two surprisingly small hydraulic pumps, connected with belts to the drive shaft coming off the diesel engine, supply most of the power to the shop. These pumps are not powerful enough to run all the machines in the shop at the same time, but that's OK because there aren't enough people to run them all either. The situation is not much different in an electrically powered shop: if you turned on all the machines at the same time, you'd likely trip a circuit breaker. Same at home; if you turn on a toaster and a hair dryer plugged into the same circuit, you'll trip the breaker.

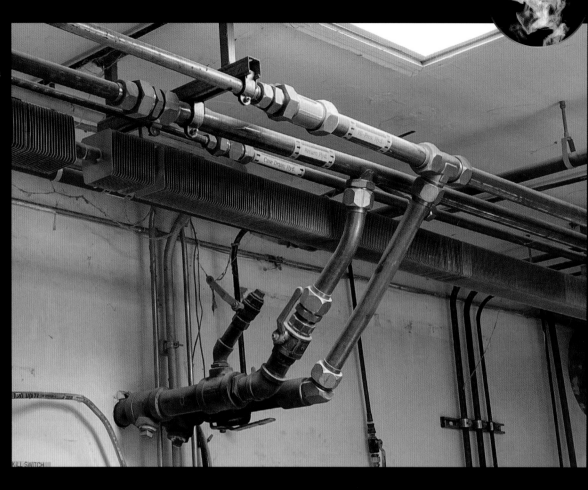

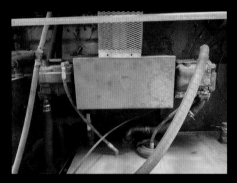

△ A network of overhead hydraulic lines distributes high-pressure fluid from the pumps to the motors and returns low-pressure fluid back to the pumps to be recirculated. This is one area where an electric shop has an advantage: running electric wires is a lot easier than plumbing in these thick steel pipes (which are much thicker than typical water pipes, due to the high pressure of the hydraulic fluid).

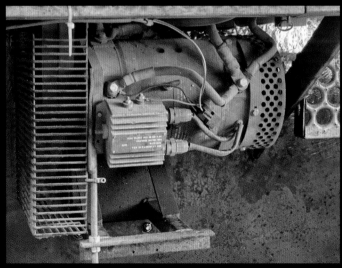

△ A small alternator (electric generator), also connected by belt to the drive shaft, supplies the relatively modest amount of electricity needed by the shop.

These Amish workshops are wonderful places, and it is a privilege to be alive to see them. They are not museums or tourist attractions; they are efficient, modern industrial shops that produce most of the furniture, kitchen cabinetry, and prefabricated construction materials made in this area. They just don't use electricity, for well-considered reasons.

Stirling Engines

THE STEAM ENGINES we've been talking about are examples of *external* combustion engines. Their fuel burns out in the open. We will soon be moving on to the much larger world of *internal* combustion engines, but we're not done yet with external combustion. There's still one really nice and one really dumb variation to look at.

Stirling engines are a bit like steam engines except they use air instead of water as their working medium. So instead of using a fire to heat up water that expands into steam, they use nearly any source of heat—from a fire to a hot cup of coffee—to heat up air, causing it to expand into...air that takes up more volume.

The remarkable thing about Stirling engines is that they can run on even a very small temperature difference between two surfaces. As long as you have one place where heat is being supplied (a heat *source*), and another place where heat can be drawn away (a heat *sink*), you can probably run a Stirling engine in the space between them. This includes running a Stirling engine on top of a cup of coffee.

If you set out a cup of coffee uncovered, heat will escape quickly, and the coffee will be cold in just a few minutes. Most of the heat is actually lost through the evaporation of water. If you put a lid

on the cup, evaporation is stopped, and heat will travel more slowly through the lid, keeping the coffee hot for much longer. And if that lid happens to be a Stirling engine, you can extract mechanical work from this flow of heat. A small engine will run for a good half hour just on the heat content of one large cup of coffee, stopping only when the coffee has cooled down to the point of being too unpleasant to drink.

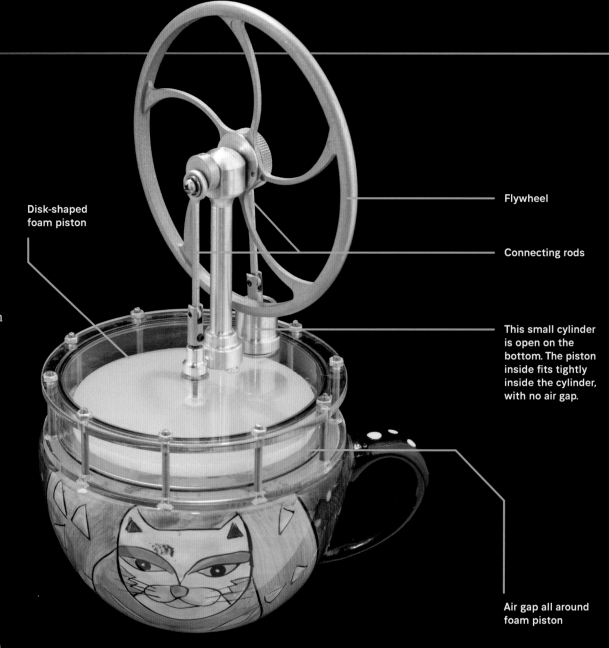

Disk-shaped foam piston

Flywheel

Connecting rods

This small cylinder is open on the bottom. The piston inside fits tightly inside the cylinder, with no air gap.

Air gap all around foam piston

△ Inside the pancake-shaped body of this Stirling engine, there is a lightweight foam disk that is a bit smaller than the hollow cylinder it sits inside (so air can easily flow around the edge of the disk). As the disk moves up and down, the air inside is pushed to opposite sides of the disk, so the air is alternately exposed to the cold upper surface and then the hot lower surface.

◁ In this position the foam disk is near the top of the pancake, so most of the air is in contact with the hot bottom. Air expands when it's heated, so the pressure inside increases, pushing up on the piston in the small cylinder. This in turn pushes on the crank arm, causing the flywheel to turn.

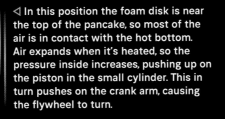

◁ As the flywheel turns, the small piston rises (it's being pushed up by the heated air), while the foam disk moves down. Air flows around the foam disk toward the cooler top surface.

◁ Now the disk is near the bottom, and most of the air is in contact with the cooler top surface. Air contracts when it's cooled, so the pressure inside drops, which pulls the small piston down. Naturally the timing has been arranged so that just as the small piston starts pulling down, its crank arm has crossed over the top, and the downward pull helps to keep the flywheel turning in the same direction.

◁ Now the foam disk is in the middle again, but this time it's moving upward while the small piston is moving down. From here we return to Step 1, and the cycle continues. All regular Stirling engines work this way, with a larger, loose-fitting piston that travels in the gap between the hot and cold surfaces, and a smaller, tight-fitting piston that extracts work from the alternately expanding and contracting volume of air.

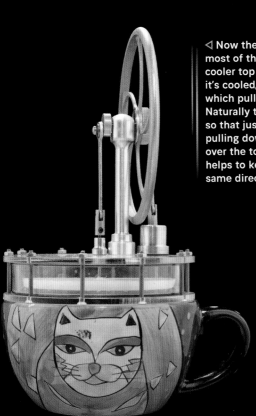

THERE ARE NOT MANY practical applications for Stirling engines. They typically generate only a very small amount of power and have very little torque (turning force). But circulating air from a wood stove is the *perfect* use case. The base resting on the surface of the stove stays hot, while radiator fins keep the top side cooler. With a good fire on, this fan spins delightfully, spreading heat around the room.

▽ Sadly for lovers of mechanical devices, and despite the fact that a woodstove fan is *the* perfect use-case for a Stirling engine, there is actually a cheaper way to make such a device. This version has a small electric motor and a solid-state thermoelectric generator, which uses the heat differential to directly generate electricity to run the motor. It costs one-fifth as much as the Stirling engine model and works just as well. Better in fact, since it doesn't need to be started manually. Oh well. I still like the mechanical version.

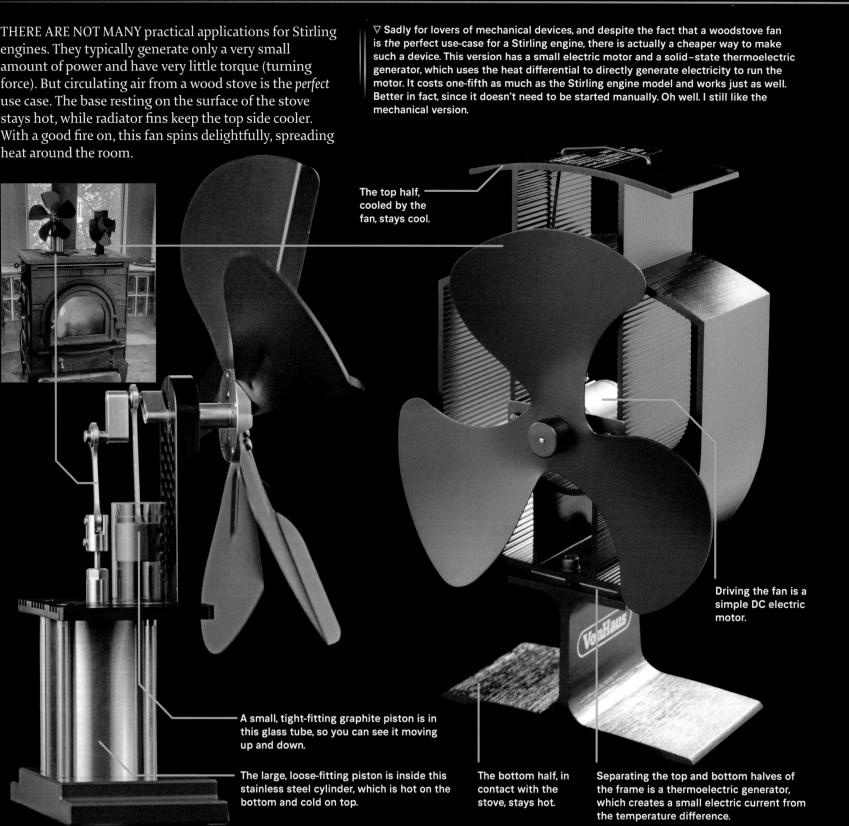

The top half, cooled by the fan, stays cool.

Driving the fan is a simple DC electric motor.

A small, tight-fitting graphite piston is in this glass tube, so you can see it moving up and down.

The large, loose-fitting piston is inside this stainless steel cylinder, which is hot on the bottom and cold on top.

The bottom half, in contact with the stove, stays hot.

Separating the top and bottom halves of the frame is a thermoelectric generator, which creates a small electric current from the temperature difference.

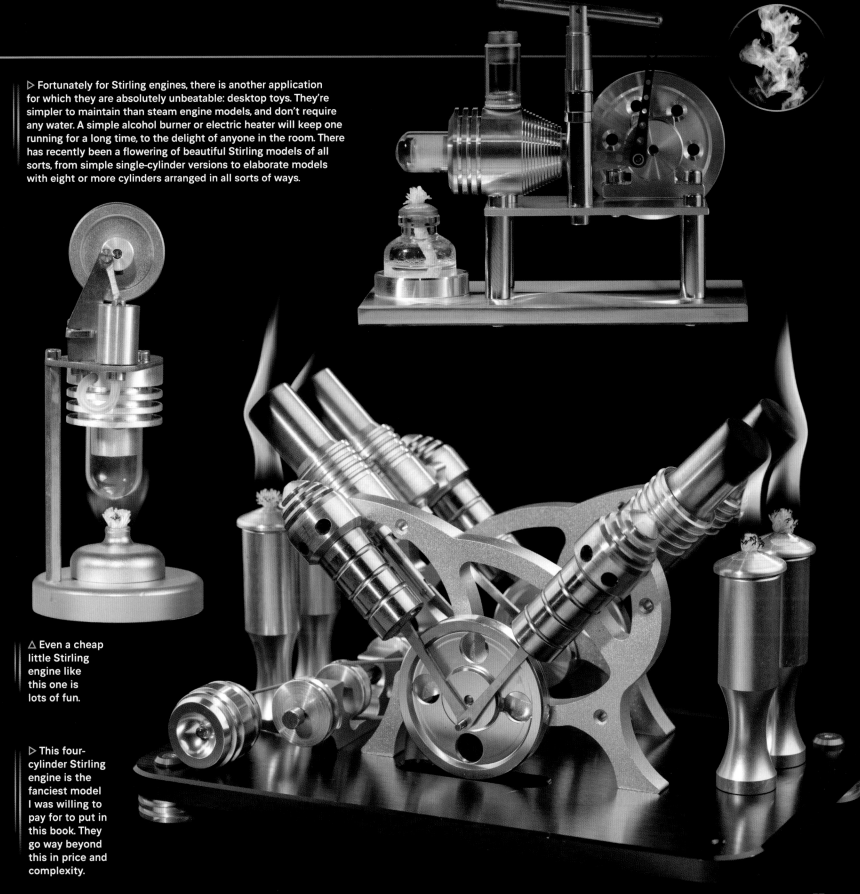

▷ Fortunately for Stirling engines, there is another application for which they are absolutely unbeatable: desktop toys. They're simpler to maintain than steam engine models, and don't require any water. A simple alcohol burner or electric heater will keep one running for a long time, to the delight of anyone in the room. There has recently been a flowering of beautiful Stirling models of all sorts, from simple single-cylinder versions to elaborate models with eight or more cylinders arranged in all sorts of ways.

△ Even a cheap little Stirling engine like this one is lots of fun.

▷ This four-cylinder Stirling engine is the fanciest model I was willing to pay for to put in this book. They go way beyond this in price and complexity.

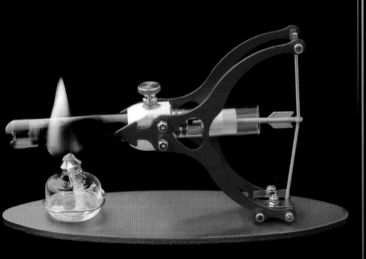

△ This is a truly baffling variation called a thermoacoustic Stirling engine. It does away with the large air-displacement cylinder (such as the foam disk in the coffee cup engine) and replaces it with...steel wool? One side of the steel wool is hot, while the other side is cool. When the engine is running, a wave of air bounces back and forth between the two sides, being alternately heated and cooled just like in a regular Stirling engine, but with no physical piston pushing it back and forth. Yes, as unlikely as it seems, these things really do work—I've seen it with my own eyes.

△ This model takes simplicity to the next level. If you don't care about extracting work from the engine, the only function of the flywheel is to store enough energy to keep the piston moving during the times when it's not being pushed or pulled by air pressure. Well then, who needs a flywheel? A rubber band stores energy to push the piston back in the opposite direction at the end of its stroke. With the alcohol burner lit, it just sits there vibrating back and forth, doing absolutely nothing useful. Perfect.

The Worst Engine Ever Made

THIS IS NOT a steam engine, it's not a Stirling engine, and it's not an internal combustion engine. Of all the engine designs ever seriously considered by anyone, it is undoubtedly the worst. The only good thing about it is its name: Flame Licker. This comes from the first step in its cycle, shown here, where a port opens up and the piston, moving away from the open flame, sucks in some burning gases. When the piston reaches the far end, the valve closes, and the gases stop burning, cool down, contract, and create a partial vacuum inside the cylinder. Air pressure on the outside of the cylinder pushes it back toward the flame, and when it's moved about halfway, the valve opens again, allowing the cooled gases to escape. Then the piston turns around and takes another gulp of flame to repeat the cycle.

The force driving the flywheel comes entirely from the atmospheric pressure on the outside pushing against the vacuum created inside the cylinder during the second stage of movement. Needless to say, the vast majority of the energy in the flame never even contributes to the operation of the engine. It just blows away in the wind.

So now that we've seen the worst engine, let's move on to a design that has been so successful it has fundamentally changed our world, has had a larger impact on culture and our view of the world than any other engine, and has easily earned its status as an engine that deserves an entire chapter in this book.

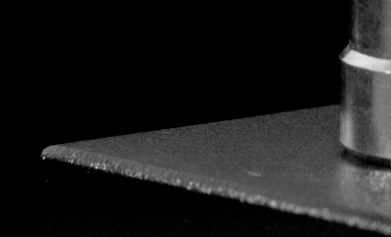

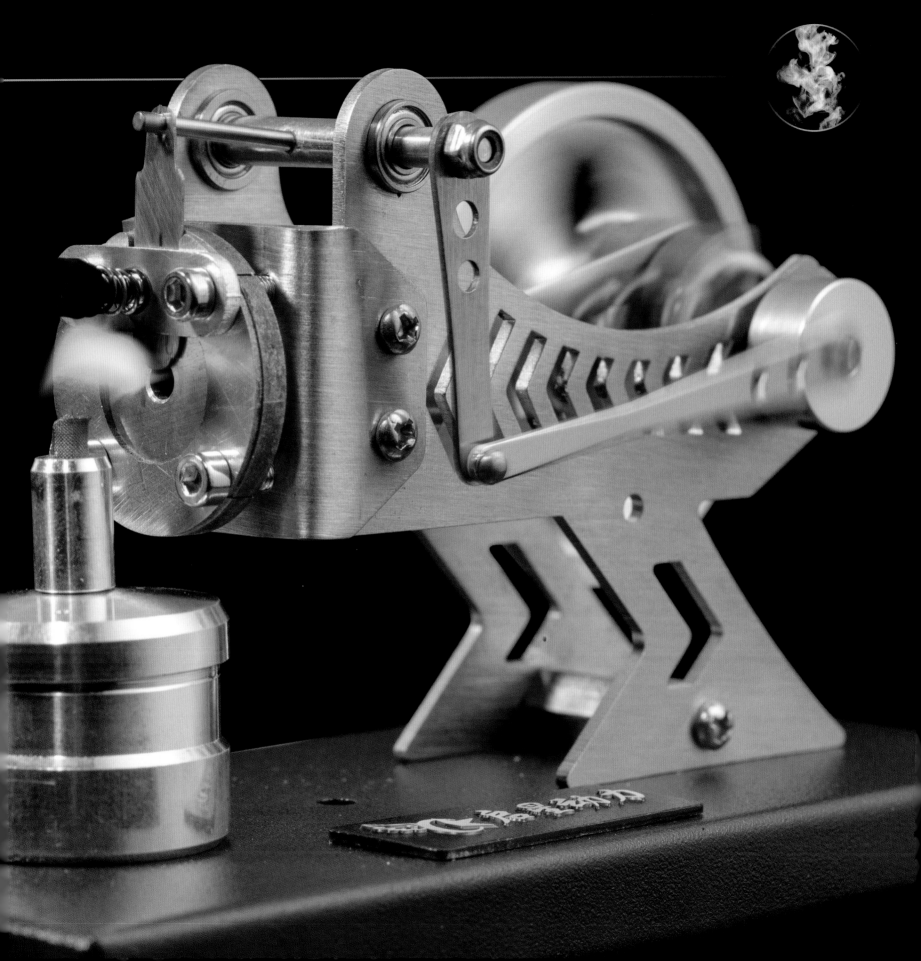

▷ Nothing tells you more about a culture than its art. This sophisticated piece, available at finer gas stations nationwide, speaks to the deep love we have for the freedom given us by motor vehicles.

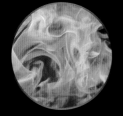 Internal Combustion Engines

FOR THE BETTER part of six generations, one kind of engine has dominated the road. Nearly every vehicle on the open road since about 1910 has run on an internal combustion piston engine. Sure, there are exceptions, and electric motors are starting to make serious inroads, but it was the internal combustion engine that changed the world, changed our relationship to distance, and changed our lives more than any other kind of motor, hands down, bar none, and it's not even close.

Internal combustion engines can be either two-stroke or four-stroke, meaning the engine completes an operational cycle in either one or two full revolutions of the crankshaft. And they can ignite their fuel using either a spark plug for engines that run on gasoline, or through the Diesel effect for

engines that run, obviously, on diesel fuel. That makes for a total of four variations: two-stroke gasoline, four-stroke gasoline, two-stroke diesel, and four-stroke diesel. Within each category, the basic operation of all piston engines that run on flammable liquids or gases is nearly identical.

The basic design of these types of engines has remained so constant that an auto mechanic today would immediately recognize all the working parts of any such engine built in any year since the 1890s. The reverse, however, is not true! An 1890s mechanic would be utterly baffled by the silicon microchips in today's electronic engine control systems, not to mention the many sensors and pollution control devices. Engines have slowly gotten more powerful and more efficient as hundreds of tiny improvements

have been added by generations of engineers and mechanics tinkering with every detail of their design—but not with any of the fundamental principles.

Internal combustion engines continue to be so widely used because they work so well. The fuel they run on is widely available and stores a tremendous amount of energy in a small space. Yes, some day fairly soon—maybe even by the time you're reading this book—electric cars will have taken over. They will be better in nearly every way—faster, more fun to drive, quieter, safer, less polluting, and hopefully better for the environment overall. But nothing can change the fact that it was the internal combustion engine that first gave us the freedom of movement that we now take as a birthright.

A Powerful Idea

INTERNAL COMBUSTION engines work by trapping a mixture of fuel and air in a confined space, called a cylinder, and then lighting it on fire. Hence the name *internal combustion*. The burning fuel creates pressure inside the cylinder, which pushes down hard on the piston that closes off one end of the cylinder. The piston in turn pushes on a crank, which converts the linear motion into rotation of the crankshaft and flywheel. In this sense these engines function exactly like steam engines, except they generate their own pressurized gas inside the cylinder—internally—rather than relying on pressurized steam being piped in from an external boiler.

The great advantage of internal combustion is that it captures not only the heat energy released by the burning fuel, but also the energy available from the expansion of the large volume of gas that's created. A small volume of liquid gasoline will turn into a large volume of exhaust gas—many hundreds of times larger. Steam engines completely waste the potential of this expansion.

Internal combustion engines exist in a vast array of sizes. Even just among the engines I've been able to drag into our studio to photograph there's a tremendous range, and out in the wild these things get pretty ridiculous.

Let's look at some of these beautiful machines and examine more closely how they work.

Gas versus Gasoline

In American English it's conventional to use the word "gas" to refer both to matter in the gaseous state (for example, air), and to a certain fraction of liquid petroleum known in the US as gasoline and elsewhere in the English-speaking world as petrol. This can cause some confusion. I will always use the word "gasoline" when referring to liquid petroleum. Anytime I say *gas*, I'm talking about that thing which isn't a solid or a liquid. To further confuse things, engines that run on gasoline can also—with minor modifications—run on an actual gas such as propane.

△ Here are the pistons and connecting rods from the smallest and largest engines I could find to photograph in our studio. The smallest is from a model airplane engine with a cylinder volume (displacement) of 0.049 cubic inches (0.8 cubic cm). It's much smaller than the head of one of the bolts that holds together the two halves of the connecting rod bearing of one cylinder of the largest engine I could get my hands on, a stationary gasoline engine with a displacement of 780 cubic inches (12.8 liters). To be honest, this connecting rod and piston are the only parts of that engine I could get through the door. The rest of the beast, courtesy of my friend Donald, is in my shed in pieces weighing several tons each.

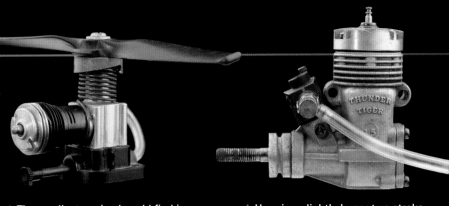

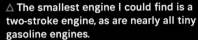

△ The smallest engine I could find is a two-stroke engine, as are nearly all tiny gasoline engines.

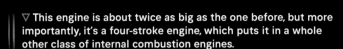

△ Here is a slightly larger two-stroke model airplane engine.

▽ This engine is about twice as big as the one before, but more importantly, it's a four-stroke engine, which puts it in a whole other class of internal combustion engines.

△ This slightly larger two-stroke engine might power a Weedwacker or another handheld gardening tool.

△ Rusty lawn mower engines are hateful. They say that for an engine to run, it needs three things: fuel, air, and spark. Lawn mower engines actually need four things: fuel, air, spark, and a new engine.

It's the two vertical push rods that give this engine away as a four-stroke engine.

◁ If you think gasoline-powered lawn mowers are miserable—and I do—this thing is ten times worse. Handheld drill/drivers need high torque at low speeds, which gasoline engines are terrible at delivering. And they want to be small and light, which this gasoline-powered drill absolutely is not. Before the invention of reasonably high-capacity rechargeable batteries, people were willing to stretch extension cords for many hundreds of feet if it meant being able to use an AC-powered electric drill instead of this ridiculous thing. You can tell from the fact that it's rusted to ruin that I haven't used this terrible drill for a very, very long time.

▽ The best kind of small gasoline engine is the kind you keep working mainly for the joy of watching it run. This antique hit-and-miss gasoline engine is running an ice cream maker at a cheese festival in one of the Amish-majority towns near me.

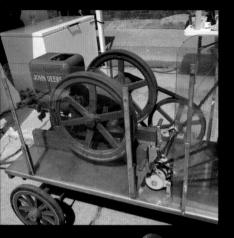

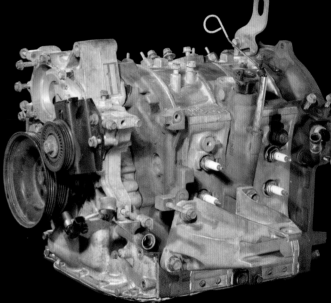

△ This forklift is carrying a load of acrylic that I am picking up to make the mechanical models you see in this book. Many of the forklifts used in industrial warehouses, including this one, are powered by four-stroke internal combustion engines, except they run on propane fuel instead of gasoline. Propane burns much more cleanly than gasoline or diesel fuel. It would *not* be a good idea to run a gasoline engine for any length of time in an indoor warehouse.

▽ Rotary engines are a peculiar variation on an internal combustion engine that runs on gasoline ignited by spark plugs, but with no pistons. Instead, fuel ignited inside the crescent-shaped cavities pushes the triangular shape around and around relatively smoothly—a good idea doomed by poor fuel efficiency and high emissions of polluting gases. They just couldn't figure out how to make it clean enough, and the design faded from use.

▽ This is a Wankel rotary engine from a Mazda car. It's physically smaller, but more powerful, than the Honda Goldwing engine.

△ Several years ago I almost satisfied a midlife crisis with a red convertible up for bid at an auction. Fortunately, the bidding went over $9,000, which was just too much given that I didn't expect my crisis to last more than a couple of weeks. But, while working on this book, I found this excellent substitute and won the bidding at a mere $350. Crisis mostly averted. It is a classic Honda Goldwing GL1000 motorcycle with a 1,000 cc horizontally opposed four-cylinder engine, which means it's really w—i—d—e, with two cylinders on each side pointing toward each other. When it was new, around 1980, it was one of the largest commercially available touring motorcycles, with an engine larger than is strictly necessary in a motorcycle.

▷ This is called a radial engine, which is not to be confused with a rotary engine. Radial engines of the kind shown here, while they look very different, are actually perfectly ordinary piston engines, with one exception: All the connecting rods meet in the middle on a single crank arm. This greatly reduces the size and weight of the engine, since there is no long crank shaft with separate crank arms for each cylinder. These were widely used on airplanes before jet engines took over.

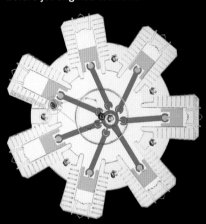

◁ This model of a radial engine shows how the pistons all connect together to a single "spider" in the center. The pistons fire in a circular pattern, following the spider around and contributing their individual pushes to the effort. Notice that one of the connecting rods—the one pointing downward and to the right—has a solid connection to the spider, rather than a rotating bearing. That keeps the spider pointing in the same direction. Without it, the force of the pistons would try to rotate the spider, rather than forcing it to push the crankshaft around.

▽ Taking apart a real engine is absolutely the best way to learn how it works. It's probably best to start with one that is beyond hope, because the chances of you getting it back together in working order the first time are slim to none. Fortunately, real engines that would normally be worth thousands of dollars are commonly available (in nonworking condition) for next to nothing at junkyards (sorry—the polite term is "auto recycling facilities"). This is a V-6, which means it has a total of six cylinders arranged in two rows of three. The rows are angled to form a V shape, which gives the engine overall a compact shape: not too long, not too wide, and not too tall.

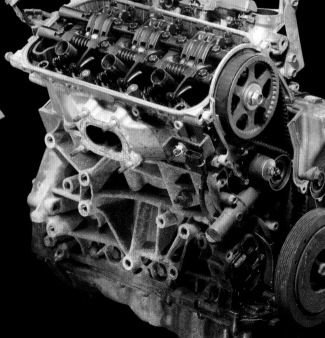

△ This engine is a step up in both size and power from the motorcycle or rotary engine. It used to be in the Acura I inherited from my dad. Sadly, a few years after he died, this engine followed him to the grave. It is a classic V-6, 24-valve design. It may look like a mess to you now, but I hope that by the end of this chapter you will be able to look back at these pictures and recognize old friends instead of strange hunks of metal. I've removed the valve covers so you can see more of the lovely bits inside.

▽ Here I've put the crankshaft, connecting rods, and pistons from the Acura engine on the prior page on a stand, arranged roughly how they would be inside the engine. Many miles has this crankshaft traveled to reach its final resting place on an acrylic stand in our studio. Each of these cylinders has had maybe half a billion explosions go off over its head, and three billion total for the engine before the last one that ended its life.

△ Eighteen-wheelers are the workhorse of long-haul trucking. They get their name from the fact that they literally have eighteen wheels: sixteen (in two groups of two axles of four wheels each) holding up the trailer, and two in front holding up an engine like this one. Large trucks typically use diesel engines because they're cheaper to operate and more reliable, but only above a few hundred horsepower.

▽ Engines scale up in two obvious ways: You can either make each cylinder larger, or you can add more cylinders. This one does both. It has twenty cylinders, compared to eight or fewer for cars. The total volume of all the cylinders—the "displacement" of the engine—is 105 liters, compared to between 2 and 5 liters for a typical car. What do you get with an engine this big? About 6,000 horsepower (4.5 megawatts). What do you do with such a large engine? This one is connected to a generator that supplies electricity to remote communities, or for emergencies when the regular power fails.

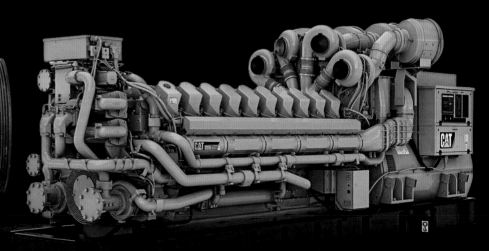

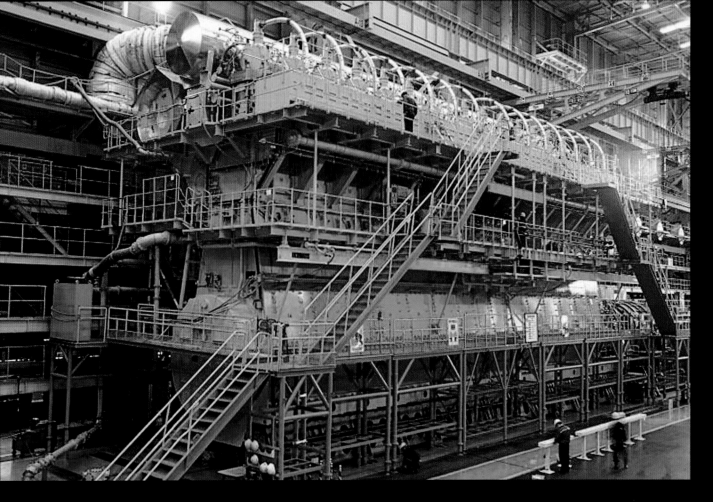

TRUCKS AND TRAINS have big engines, but to find a really *huge* engine you have to look to the sea. Container ships, among the largest moving objects on earth, have engines the size not of houses, but of three- or four-story apartment buildings. This fourteen-cylinder Wärtsilä-Sulzer RTA96-C marine diesel is 44 feet (13.5 m) tall and 87 feet (26.5 m) long. It generates 109,000 horsepower at full power.

Each cylinder is over 3 feet (1 m) in diameter and moves just over 8 feet (2.5 m) from the top to the bottom of each stroke. Picture something that big, weighing 12,000 pounds (5.5 metric tons), moving from floor to ceiling and back in less than a second. That, by the way, is just the weight of one cylinder. The crankshaft is 660,000 pounds (300 metric tons) and the whole engine weighs about 5 million pounds (2,300 metric tons).

This is clearly a big engine, but the statistic that really blows my mind is the amount of fuel it burns. On *each stroke* of *each cylinder*, about 5.5 ounces (155 g) of heavy marine fuel oil (bunker oil) is injected and burned. That's enough to drive a car about two miles, and this beast uses it up just to push one cylinder down a single time.

Running at a typical operating speed of 100 revolutions per minute under full load, the engine uses about one gallon (3.8 L) of fuel, per *second*. Imagine if your car went through a full tank of gasoline every 15 seconds! Its typical daily fuel consumption is 250 tons (225 metric tons). At normal cruising speed this works out to about 0.01 miles per gallon (.0042 km/L). Put another way, it gets about 50 feet per gallon (4 m/L). Sounds bad, but on the other hand, considering the absolutely vast amount of cargo it is carrying, this is still by far the most efficient means of moving large amounts of stuff.

Four-Stroke Engines

ALTHOUGH FOUR-STROKE ENGINES are mechanically more complex than two-stroke versions, I find them easier to understand, because the various stages of their operation are better separated. So we'll start with them.

To make a continuously running engine, each cylinder needs to go through a multistage cycle: It needs to be filled with fuel and air, which then needs to be compressed, ignited, and allowed to expand. After the expansion is completed, the exhaust gases need to be flushed out to make room for a new charge of fresh air and fuel.

Four-stroke engines get their name from the fact each of these four operations—intake, compression, combustion, and exhaust—is carried out during one "stroke" of the piston, meaning movement of the piston from one end of the cylinder to the other, either from top to bottom or bottom to top. In a complete four-stroke cycle, the piston moves up, down, up, down, meaning the crankshaft makes two full revolutions.

The diagrams on the following pages apply equally to gasoline engines and diesel engines. The only difference between them is the way in which the fuel is ignited. In gasoline engines a spark plug lights the fuel at just the right moment, while in diesel engines the compression of the air in the cylinder makes it so hot that when fuel is injected at the right moment, it catches fire spontaneously.

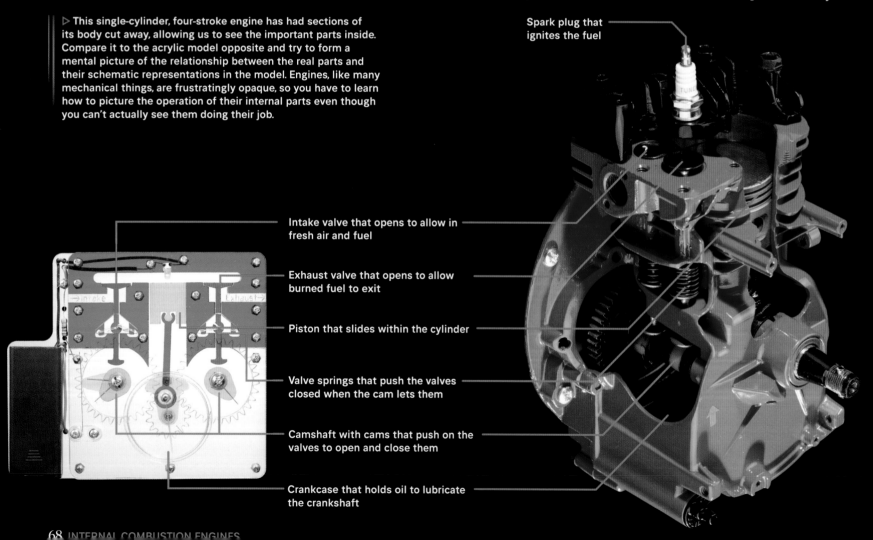

▷ This single-cylinder, four-stroke engine has had sections of its body cut away, allowing us to see the important parts inside. Compare it to the acrylic model opposite and try to form a mental picture of the relationship between the real parts and their schematic representations in the model. Engines, like many mechanical things, are frustratingly opaque, so you have to learn how to picture the operation of their internal parts even though you can't actually see them doing their job.

Spark plug that ignites the fuel

Intake valve that opens to allow in fresh air and fuel

Exhaust valve that opens to allow burned fuel to exit

Piston that slides within the cylinder

Valve springs that push the valves closed when the cam lets them

Camshaft with cams that push on the valves to open and close them

Crankcase that holds oil to lubricate the crankshaft

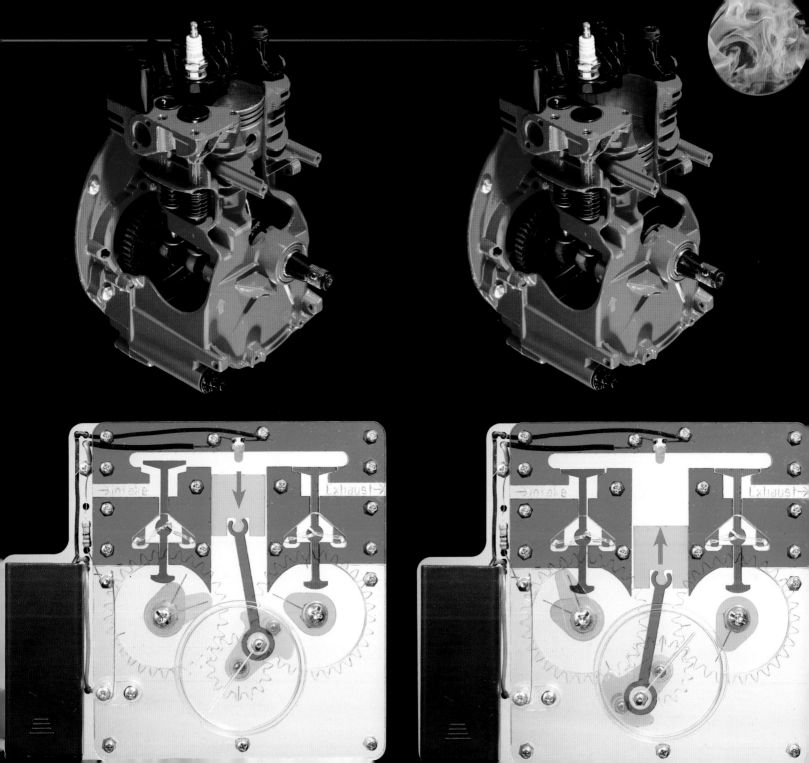

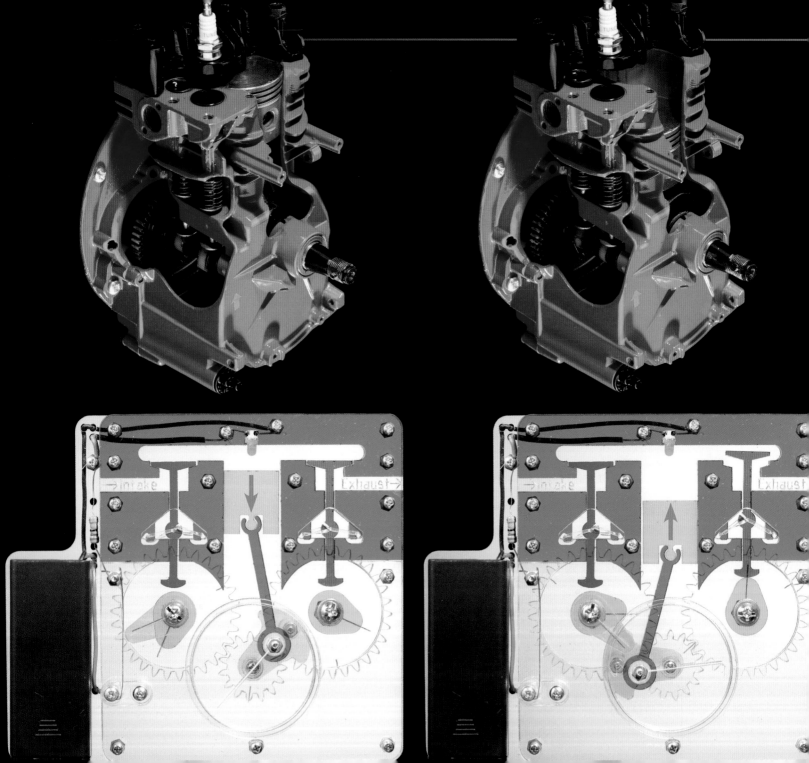

THERE ARE THREE common ways of arranging the valves in a four-stroke engine. I'm showing them here in model form so you can appreciate the geometric relationships between the parts. In real engines, of course, this is all metal, much more compact, and overlaid in confusing three-dimensional ways.

Flathead model ▽

This is called a flathead engine because there's basically nothing above the cylinder except the spark plug. These engines are simple and compact but perform poorly because the horizontal space connecting the cylinder and valves limits the amount of compression that can be achieved. It also makes the flow of gases inefficient: for good performance, you want the valves as close as possible to the top of the piston.

This space makes the engine inefficient.

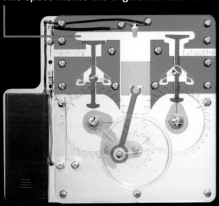

Pushrod design ▷

This pushrod engine combines the efficient valve placement of the overhead cam design with the compact timing gears of the flathead design. But the speed of the engine is limited by the fact that a lot more weight has to move back and forth with each opening of the valves. In the other designs only the valve itself moves, but here it's the valve plus the rocker arm, pushrod, and lifter. This might seem like a poor compromise, but for engines that don't need to operate at very high revolutions per minute (rpm), including most passenger car engines, it's actually the best and most common solution.

Dual overhead cam design ▷

This overhead cam engine puts the valves where they should be, directly above the cylinder. The price you pay is that the cams are now literally on the opposite side of the engine from the crankshaft that needs to drive them. This means the engine needs to be taller and requires a toothed belt (called the timing belt or chain), or a long gear linkage, to connect the crankshaft and the camshafts.

▷ The timing belt or timing chain must maintain perfect synchronization between the crankshaft and the camshafts. If it slips even one tooth the engine will not run, or it will run so badly that you don't want it to run. In some designs the engine will actually destroy itself if the belt breaks, because the piston will run into the valves if they are not raised (closed) every time the piston comes to the top.

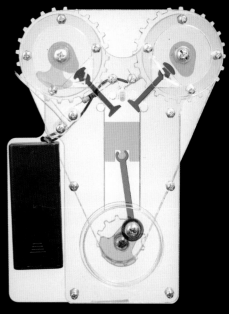

Pushrods transfer force from cam to rocker arm.

Rocker arms reverse the direction of the cam's motion, turning a push up from the cam into a push down on the valve.

Lifters follow the cams up and down.

The timing gears in this type of engine can be small and close together. In real engines they are much smaller, compared to the size of the cylinder, than in this model.

Guess the Engine

NOW THAT YOU'RE familiar with these different types of engines, it's fun to look at an encased engine and try to figure out what's inside. The presence of pushrods, for example, means there must be rocker arms inside, and sure enough, when you take off the cover, there they are.

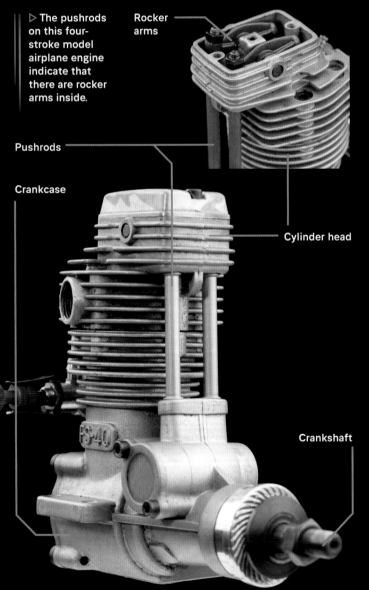

▷ The pushrods on this four-stroke model airplane engine indicate that there are rocker arms inside.

Rocker arms

Pushrods

Crankcase

Cylinder head

Crankshaft

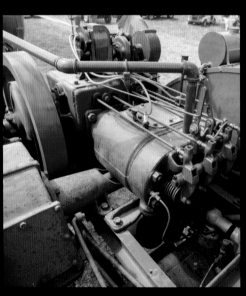

△ Antique engines often have exposed rocker arms. These things operate at very low speeds, so you can see the valves opening and closing a few times per second. The sound is beautiful—a complex, multilayered rhythm.

◁ I stared at this aircraft engine in the Henry Ford Museum in Dearborn, Michigan, for a long time. It's obviously a four-stroke engine with pushrods and rocker arms, but I was perplexed by the fact that there's only one pushrod per cylinder—but two valve springs? How does this make any sense? I finally figured out that there are two pushrods, as you'd expect, but one is a hollow tube and the other is inside that tube. What looks like a single pushrod is actually two concentric pushrods. Almost fooled me!

▷ Valves are opened by the positive action of the cam pushing down on them. But when it's time for the valve to close again, all the closing force is supplied by the valve springs. In many engine designs the strength of the valve springs is a key factor limiting the maximum speed at which the engine can turn. Doubling up the springs, one inside another, lets you get more force and more reliability—a broken valve spring can easily mean a destroyed engine, and certainly means a lost race. The ultra-high-quality steel and careful preparation and quality control mean a single one of these springs can cost hundreds of dollars. (Not this one; it's a cheap brand.)

The Intake Stroke

FOUR-STROKE ENGINES have four phases of operation, each corresponding to the movement of the piston from top to bottom, or bottom to top. The four strokes repeat endlessly, so there isn't really a "first" stroke with which to start an explanation of the sequence, but the intake stroke is perhaps the most logical place to dive in. At this point in the cycle, we have just finished the exhaust stroke and the cylinder is at the top, filled at minimum volume with a small amount of leftover exhaust gas. We say

that the cylinder is at the "top dead center" position—"top" because it's at the top, "dead" because this is the point at which the piston's movement momentarily stops as it changes direction from moving up to moving down, and "center" because in this position the connecting rod and crank arm are centered directly below the piston.

At the start of the intake stroke, the cylinder begins moving down, which creates suction. The intake valve opens as soon as the cylinder starts moving, allowing air to

be sucked in. Older gasoline engines use a "carburetor" to mix a fine mist of gasoline with the air just before it gets sucked into the cylinder. Newer engines use fuel injectors to spray liquid gasoline directly into the cylinder as it is being filled with air. This allows for precise, independent control of the amount of air and fuel entering the cylinder, making the engine more efficient, cleaner-burning, and more powerful. All diesel engines use fuel injection, and always have; it is a fundamental part of the Diesel design.

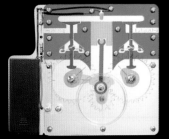
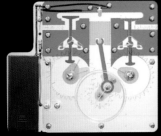

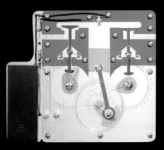
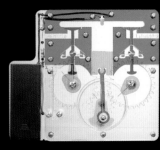

△ Both valves closed

△ Intake valve open

△ Piston traveling down, and sucking in air and fuel

△ Piston still traveling down, and sucking in air and fuel

△ Intake valve closed again

Carbureted vs. Fuel-Injected Engines

THE MORE FUEL and air you feed into the cylinder, the more powerful an explosion you will get when it's ignited, and the more power the engine will deliver. A tricky detail here is that if you want an efficient, clean burn, you need to vary both the amount of air and the amount of fuel. If they are too far out of balance, you will get either incomplete combustion if there's too much fuel and not enough air, called a rich mix; or excess generation of noxious, polluting gases if there's too much air and not enough fuel, called a lean mix.

In older cars the mixture of fuel and air is regulated with a carburetor, a name that derives from the fact that it is supplying carbon-containing fuel. This device combines three key components. In the front is the choke, a vane that can be rotated to partially block the flow of air into the engine from the outside *before* fuel has been mixed in. In the middle is a small hole from which gasoline can be sucked in from a pool underneath. Finally, on the back is another vane, the throttle, that regulates the flow of air *after* fuel has been added.

When the piston retreats during the intake phase, it creates suction, trying to pull air through the carburetor. If the throttle is closed or almost closed, nothing much is going to enter the cylinder, and the engine will run slow. If the throttle is open, the suction reaches what is called the fuel inlet stage. If the choke in front is closed, not much air can flow in, so the suction gulps up a lot of fuel and not much air, resulting in a rich fuel mixture, which is what you need to get the engine started. If the choke is open (the normal running position), you get mostly air and less fuel. The fuel feeding mechanism is designed so that with a wide-open choke, the fuel mixture will be correct for a warmed-up engine.

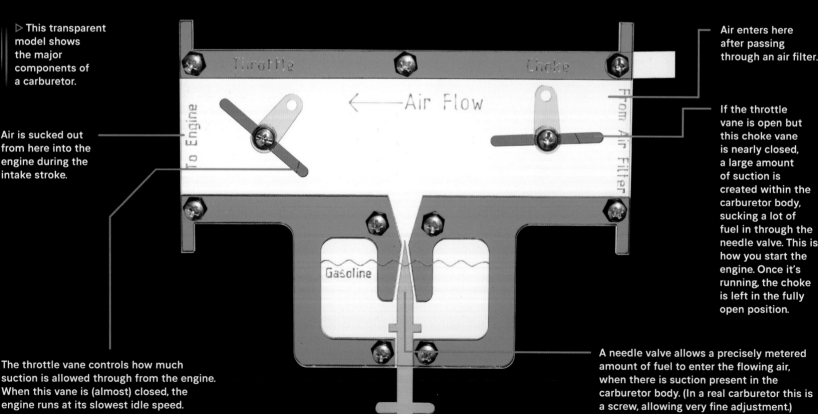

▷ This transparent model shows the major components of a carburetor.

Air is sucked out from here into the engine during the intake stroke.

The throttle vane controls how much suction is allowed through from the engine. When this vane is (almost) closed, the engine runs at its slowest idle speed.

Air enters here after passing through an air filter.

If the throttle vane is open but this choke vane is nearly closed, a large amount of suction is created within the carburetor body, sucking a lot of fuel in through the needle valve. This is how you start the engine. Once it's running, the choke is left in the fully open position.

A needle valve allows a precisely metered amount of fuel to enter the flowing air, when there is suction present in the carburetor body. (In a real carburetor this is a screw, allowing very fine adjustment.)

Throttle

Choke

← Air Flow

To Engine

From Air Filter

Gasoline

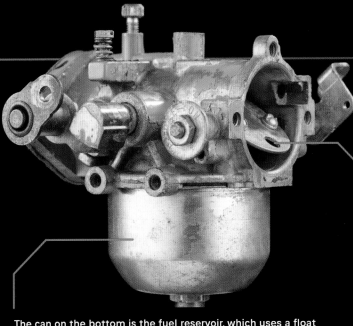

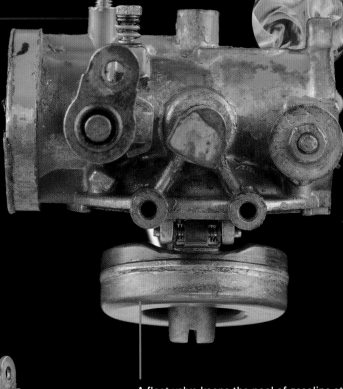

The can on the bottom is the fuel reservoir, which uses a float valve to maintain a constant level of gasoline (like the float valve in a toilet tank).

△ A carburetor is an inscrutable thing: one of those strange barnacles attached to the side of an engine that makes no sense to the uninitiated. This one came from an old generator. Looking in from the right side we see the choke vane. When it's rotated to a vertical position it blocks nearly all the incoming air, while in a horizontal position air flows freely.

Looking in from the left side we see the throttle vane, which can block the flow of both air and fuel.

There's no small hole in the throttle vane like there is in the choke. Instead, this screw allows you to limit how closed-off the vane can get. This is the idle adjustment, which sets how slowly the engine runs at its minimum speed.

This needle valve regulates the flow of gasoline into the carburetor. It is adjusted so that the engine runs properly with the throttle and choke open.

The needle is just visible when the valve is closed all the way.

Rotating this handle turns the needle screw, moving the needle in and out by a very precise amount, allowing very fine control of fuel flow.

This small hole ensures that the choke always allows a certain minimum amount of air in; otherwise the engine would flood with gasoline and stop.

A float valve keeps the pool of gasoline at a constant level.

NEWER ENGINES often use fuel injection instead of a carburetor. There is still a throttle to control the amount of air entering the cylinder, but there is no choke or fuel adjustment needle valve. Instead, fuel is squirted directly into the cylinder through one of these injectors, which fits into a hole in the cylinder wall.

▷ In a fuel-injected engine, a computer decides how much fuel to send to each cylinder based on the throttle position and the temperature of the engine.

Fuel sprays into the cylinder from these tiny holes.

In this section there is a small solenoid that drives a piston pump to drive fuel out the holes and into the cylinder.

The fuel comes through hoses connected to a fuel pump.

Electrical signals from the computer tell the injector how much fuel to give each time.

Adding Oxygen

THE MAXIMUM AMOUNT of power you can get from an engine (its horsepower) depends on how many cylinders it has and how large those cylinders are. But adding cylinders or making them bigger also makes the engine bigger and heavier, which is not ideal if you're building a race car, for example.

Another way to increase horsepower is to pack more fuel and air into each cylinder. Adding fuel is easy: You can either adjust the carburetor or inject more through the fuel injectors. But neither

does any good unless there's a corresponding amount of air to burn the fuel. It's a lot harder to add air, because to do so you have to increase its pressure, which requires an air compressor of some sort. Enter the turbocharger or supercharger, both of which are used to pump extra pressurized air directly into the cylinders. (The difference between them is that a turbocharger is powered by a turbine that spins in the exhaust gas stream, while a supercharger is powered mechanically by a belt connected to the crankshaft.)

▷ Ordinary air is only about 22 percent oxygen; the other 78 percent is almost all nitrogen, which just gets in the way. Nitrogen doesn't burn and it cools down the reaction, robbing the engine of power. If you were to replace the nitrogen with more oxygen, you would *vastly* increase the power of the engine. You could also blow it sky-high—a mixture of pure oxygen with fuel is a powerful explosive. And besides, there is a much more convenient and safer way to add oxygen to the mixture: build it right into the fuel molecules.

▽ This is a supercharger (which means it's powered by the crankshaft) on a 1935 Auburn Boattail Speedster. You can tell from the shiny chrome exhaust pipes that this was a pretty fancy car! Like sports cars today, it was all about going fast and looking cool.

△ One popular choice for adding extra oxygen to high-power model engines and drag racers is a mixture of methanol (wood alcohol, CH_3OH) and between 10 percent and 90 percent nitromethane (CH_3NO_2). The *nitro* in the name of this substance is a clue to its explosiveness: you find the same root in such things as "nitrocellulose" (guncotton) and "nitroglycerine" (the high explosive that gives dynamite its kick). Nitro groups (NO_2) do two things: they supply built-in oxygen to the fuel, allowing it to burn itself up without needing as much (if any) air; and they include nitrogen, which turns into N_2 gas, creating a larger volume of gas and thus a higher pressure in the cylinder. Nitro fuel is almost, but not quite, an explosive.

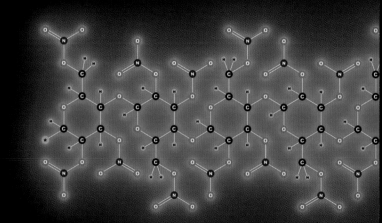

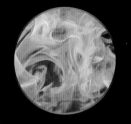

▽ Methanol

▽ Nitromethane

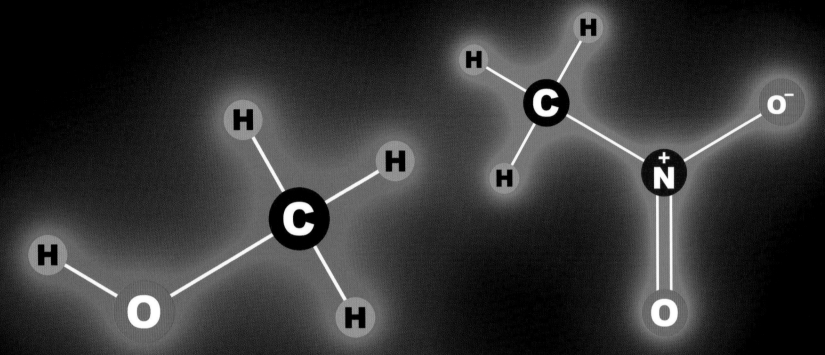

▽ Nitroglycerine

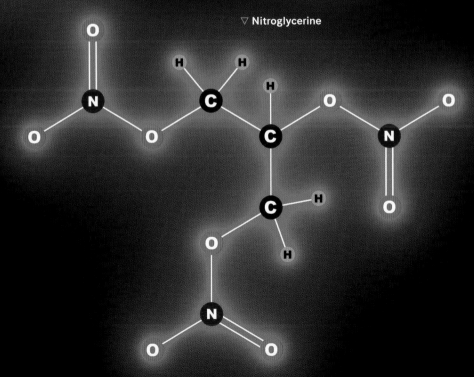

▽ Nitrocellulose Polymer

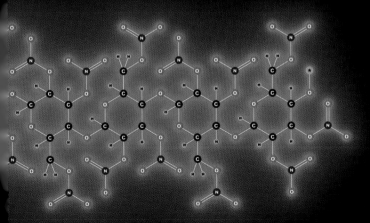

The Compression Stroke

AFTER THE INTAKE stroke comes the compression stroke. During this phase both intake and exhaust valves are closed, sealing off the cylinder from the outside world. The piston travels from bottom dead center to top dead center, greatly decreasing its internal volume and compressing the air or fuel-air mixture in preparation for igniting it.

The *compression ratio* of an engine is the difference between the maximum and minimum volume of the cylinder as it closes up during the compression stroke. A "high-compression" engine squeezes the cylinder contents down to a smaller fraction of their original volume, which increases power and fuel efficiency.

It takes a *lot* of force to execute this compression. Even with a small model engine it can be difficult to turn the crankshaft manually through a compression cycle, and in a full-size engine it's impossible without some form of mechanical assistance, or some way of partially "unloading" the cylinder to reduce the amount of compression temporarily.

Why would you want to manually crank an engine through its compression stroke? Because that's how you get it started. You have to get some fuel and air into the engine, and get it compressed, in order to initiate the first combustion stroke during which you will actually get power out of the engine. In modern cars we're spoiled by deceptively small but very strong electric starter motors. Early cars had to be hand-cranked to get them going; "starter" was a job description, not a part name. Larger engines, including the huge diesel engines in ships, large stationary engines, and locomotive engines, use a variety of methods to get themselves going. Some have compressed air tanks that save up pressure from the last time the engine was run, and others have multi-hundred-horsepower gasoline engines as starter motors (and those gasoline engines in turn have electric starter motors).

△ The crank on the front of this car marks it as coming from a time when even the fanciest cars had to be started by hand. In those days a "starter" was a person, not an electric motor.

▽ Both valves remain closed for the full stroke.

▽ As the piston travels up, it compresses the fuel-air mixture.

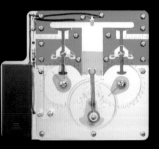
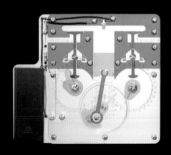
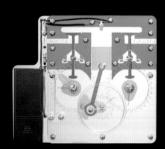
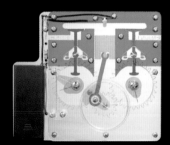
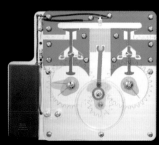

> Small 12-volt DC electric motors are used to start car engines. These motors are incredibly powerful (high torque) for their size and draw a tremendous amount of current from the battery, 200 amps or more in a typical car. That's why car batteries are optimized to deliver short bursts of very high current, unlike deep-cycle batteries that are optimized to deliver a large total amount of energy over a longer period of time.

> It takes a lot of energy to push a piston through its compression stroke. Where does that energy come from when the engine is running? In a large single-cylinder engine it is supplied by the inertia of a large, very heavy flywheel. It takes a lot of energy to get this thing going, but once it's going it also takes a lot of energy to stop it, or even slow it down. Since the piston delivers power only once every other full revolution of the engine (which will typically be running at a very slow speed), the flywheel has to store enough energy not only to overcome the next compression stroke, but also to keep supplying power to whatever the engine is connected to. That's why you see such truly massive, multi-ton flywheels on this type of engine.

In a multicylinder, high-rpm engine, the other cylinders can help push through the compression stroke of whichever cylinder is up next. The flywheel doesn't need to be nearly as large as in a single-cylinder engine, but it still helps to smooth out the uneven supply of power.

Why Some Gasoline Costs More

ONE PROBLEM WITH high-compression engines is the tendency for the fuel-air mixture to ignite prematurely while it's being compressed. This "knocking" happens because air heats when it's compressed and the more you compress it, the hotter it gets. If it gets hot enough to light the fuel before the cylinder reaches top dead center, the force of the explosion will push the cylinder down when it's supposed to be going up, in effect pushing the engine *backward*. This is bad.

The solution to knocking is to use a fuel that's more resistant to ignition by compression. For historical reasons this is measured by the "octane" rating of the fuel. A higher-octane fuel can operate without knocking in a higher-compression engine. That's why older, less efficient cars can run on cheap low-octane fuel, but whiny little gas sippers—er, sorry, I mean modern fuel-efficient cars—need to be pampered with expensive high-octane premium fuel.

The term "octane rating" comes from a particular hydrocarbon called octane—more specifically the iso-octane molecule. This molecule gets its name from the fact that it has eight (*octo* as in "octopus") carbon atoms. It happens to have a very high resistance to ignition by compression, compared to the more common heptane (7-carbon), hexane (6-carbon), and pentane (5-carbon) molecules found in gasoline. High-octane fuels may in fact contain some actual iso-octane, but more commonly they get their knock resistance from other, cheaper substances that have the same effect.

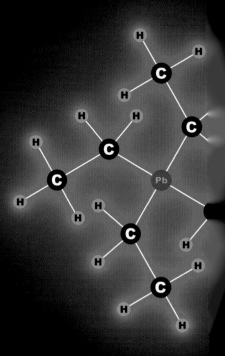

◁ The big octane number you see at the gas pump tells you how resistant the fuel is to auto-ignition when it's compressed. Period. If your car has an engine that doesn't have a high enough compression ratio to require premium fuel, there is no advantage to using it. You won't get more power or better gas mileage, just higher-priced gasoline.

▽ Iso-octane is a branched hydrocarbon with eight carbon atoms.

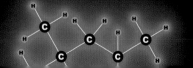

△ The most infamous substance ever used octane rating of fuel was tetraethyl lead. It on its own, and once burned in an engine i lead, which settles on, in, and around every automobiles distributed a fine dusting of l world. Lead is a chronic cumulative neurot it builds up in your body over time from ti eventually causes brain damage. There is n limit for lead exposure, and yet for decade cars to be blown into the air. Fortunately, t lead has been banned for many years alm world, other than a very few exceptions for

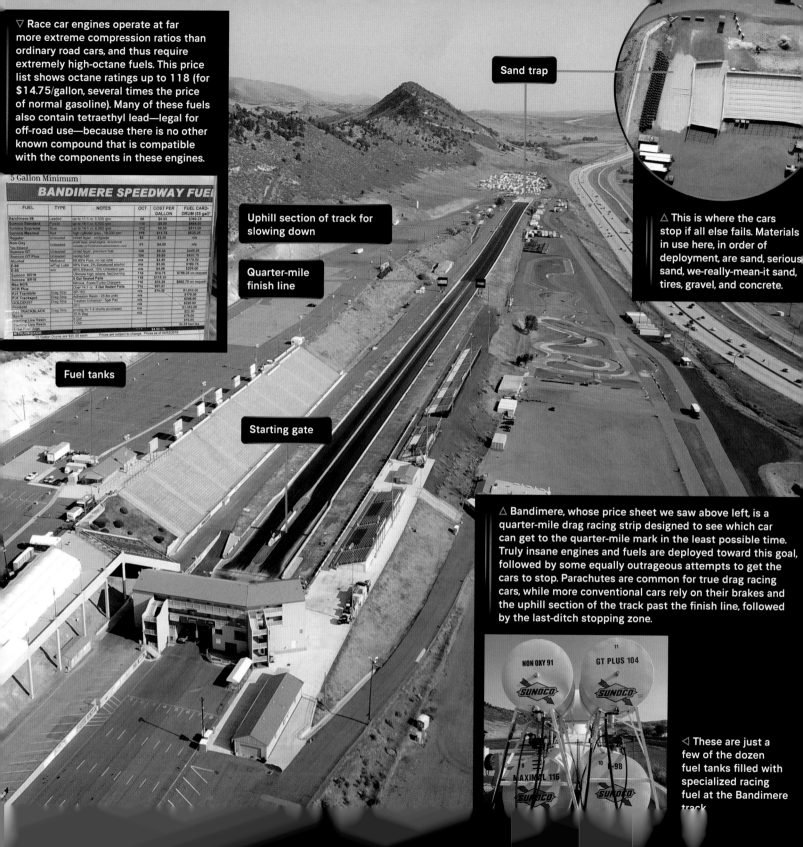

▽ Race car engines operate at far more extreme compression ratios than ordinary road cars, and thus require extremely high-octane fuels. This price list shows octane ratings up to 118 (for $14.75/gallon, several times the price of normal gasoline). Many of these fuels also contain tetraethyl lead—legal for off-road use—because there is no other known compound that is compatible with the components in these engines.

Sand trap

△ This is where the cars stop if all else fails. Materials in use here, in order of deployment, are sand, serious sand, we-really-mean-it sand, tires, gravel, and concrete.

Uphill section of track for slowing down

Quarter-mile finish line

Fuel tanks

Starting gate

△ Bandimere, whose price sheet we saw above left, is a quarter-mile drag racing strip designed to see which car can get to the quarter-mile mark in the least possible time. Truly insane engines and fuels are deployed toward this goal, followed by some equally outrageous attempts to get the cars to stop. Parachutes are common for true drag racing cars, while more conventional cars rely on their brakes and the uphill section of the track past the finish line, followed by the last-ditch stopping zone.

◁ These are just a few of the dozen fuel tanks filled with specialized racing fuel at the Bandimere track.

BANDIMERE SPEEDWAY FUEL
5 Gallon Minimum

FUEL	TYPE	NOTES	OCT	COST PER GALLON	FUEL CARD-DRUM (55 gal)*
Bandimere 98	Leaded	up to 11.1 cr, 5,500 rpm	98	$8.65	$360.25
Sunoco Standard	Purple	up to 14.1 cr, 6,500 rpm	110	$9.80	$484.00
Sunoco Supreme	Blue	up to 14.1 cr, 9,000 rpm	112	$9.50	$511.50
Sunoco Maximal	Red	high cylinder pres., 10,000 rpm	110	$11.75	$635.25
Regular	Unleaded	street legal - midgrade	85	$3.50	n/a
Non-Oxy	Unleaded	street legal, small engine, recreational vehicles (ATVs/snowmobiles/classic cars)	91	$4.00	n/a
*No Ethanol	Unleaded	street legal - premium fuel	100	$6.35	$448.25
Sunoco GT	Unleaded	racing fuel	104	$9.85	$630.75
Sunoco GT Plus	Methanol	99.90% Pure, no top lube	n/a	$3.65	$179.50
Alcohol	w/Top Lube	98% Pure, 2% Denatured alcohol	n/a	$4.00	$189.75
E-98	w/o	85% Ethanol, 15% Unleaded gas	n/a	$14.75	$209.00
E-85		Ultimate high octane, fast burning	118	$115.10	$788.25 on request
Sunoco SR18	5 Gal Sealed Pails	Nitrous, Super/Turbo Chargers	118	$16.25	$862.75 on request
Sunoco SR18		Over 14.1 cr, 5 Gal Sealed Pails	116	$87.07	
Max NOS			11+	$76.30	$1,010.00
HCR Plus	Drag Strip	SR-163	n/a		$179.00
PJ1 Trackbite	Drag Strip	Adhesion Resin - 25 lbs units	n/a		$246.00
PJ1 Trackspot	Drag Strip	Traction Enhancer, 5gal Pail	n/a		$220.00
GOLDDUST		SFC			$1,383.00
ProGold	Drag Strip	pricing for 1-4 drums purchased	n/a		$22.50
PJ1 TRACKBLACK		20 lb Bag			$76.00
Xsorb		5 Gal			$15.00
Starting Line Resin		5 Gal			31.33 Incl tax
Starting Line Resin				$4.50 / lb.	
5 Gal Fuel Jugs					
NITROUS OXIDE	NO2	55 Gallon Drums are $20.00 each. Prices are subject to change. Prices as of 06/02/2019			

NON OXY 91

GT PLUS 104

MAXIMAL 116

E-98

The Combustion Stroke

ERE WE ARE! This is it! This is the money stroke, the one that delivers what we've been waiting for: *chemical energy turned into mechanical work.*

The combustion stroke, also known as the power stroke or expansion stroke, is when the fuel is ignited and chemically combined with air to create a large volume of gas, which creates an enormous amount of pressure in the cylinder. This pushes the piston down, which pushes on the connecting rod, which pushes on the crank arm, which turns the crankshaft, which turns the gears in the transmission, and from there through the differential to the wheels (or to whatever else the engine might be connected to).

The combustion stroke is not just the most important, it's also by far the most complex phase of engine operation, because it's both mechanically complicated, and chemically and hydrodynamically messy. There's a flame traveling through a mixture of gases swirling around inside a chamber that's changing shape; you can't get much more complicated than that in terms of gas flow and reaction kinetics.

In the early days of engines, folks just

The spark goes off in the middle of a mixture of fuel and air.

△ You can see here that the cylinder is not quite at top dead center yet.

shrugged their shoulders and said, hey, it's burning and the car is moving, so good enough! But as cars began to be manufactured by the millions, and gasoline used by the shipload, we started to look more closely at just exactly what was happening inside the cylinder. Thousands of people have dedicated their professional lives to the combustion stroke. Improvements of 1 or 2 percent in combustion efficiency have a huge impact on the environment, the price of oil, and the air quality in cities.

The "explosion" is a complicated, multistage chemical reaction that needs to happen with precision in the correct amount of time and at the right temperature. Too fast, and the pressure generated will be too high, stressing the engine. Too slow, and there will be unburned fuel left when the exhaust valve opens. Too cold, and partially burned fuel molecules will be left over. Too hot, and undesirable molecules will be created by the breakdown of nitrogen in the air. Endless variations of cylinder and valve shape, fuel composition, spark plug geometry, and timing have been tried. Countless detailed computer simulations have been run to find optimal combinations of parameters. At this point the efficiency is so close to theoretical perfection that it's unlikely there will be much more improvement before the whole issue becomes moot as these engines are replaced by electric motors.

▽ Both valves remain closed for full stroke.

▽ The piston is traveling down, delivering power to the crankshaft.

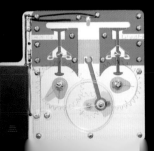
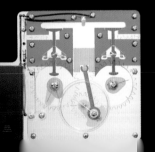
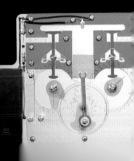

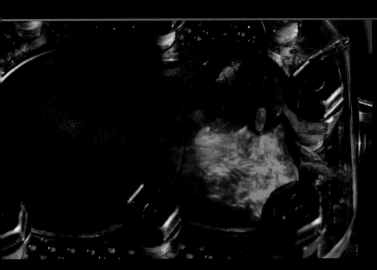

△ A fraction of a second later the cylinder has reached top dead center, and the fire has just started to spread.

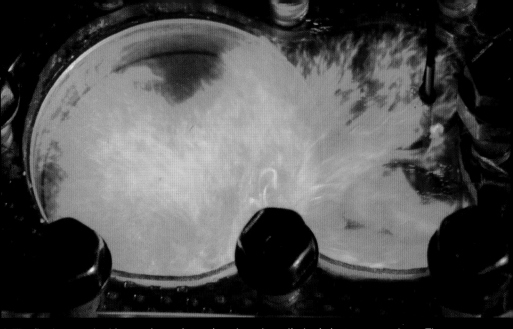

△ The fire has reached its maximum intensity when the cylinder is just partway down. The pressure inside is huge, pushing down very hard on the piston. By the time it reaches the bottom the fire will be out, and the remaining pressure inside will push everything out as soon as the exhaust valve opens.

◁ Valves have a hard life. Their heads need a large surface area to allow for a large enough opening to let gases in and out, but they also work right next to the explosion. This means they absorb a *lot* of heat, which they need to conduct into the main engine block, where it is carried away by circulating cooling water. The valves in some engines, including the famous Rolls-Royce Merlin aircraft engine of World War II, get so hot that the valve stems have to be partially filled with sodium metal, which melts and turns into a very good conductor of heat when the engine is running. The space inside the valve stem is intentionally only partially filled so the liquid metal can slosh back and forth (sort of like washing the inside of a bottle by half-filling it and then sloshing the water around).

Lighting the Fire

GASOLINE ENGINES use a spark plug to ignite the fuel-air mixture. Spark plugs have a hellish life: They spend their days at work right in the middle of dozens or even hundreds of powerful explosions occurring every second the engine is running. All kinds of burning, corrosive gases are flowing around them at very high temperatures, and if that's not enough, they are responsible for creating the 100,000°F (60,000°C) spark that starts the whole mess going.

The best spark plugs these days are made with platinum or iridium electrodes and can last a remarkable 50,000 miles (80,000 km) or more of continuous operation. During their life they create and survive something like a hundred million explosions. As with every other part of the engine, the high reliability of spark plugs is the result of decades of refinement, tinkering, and scientific research in disciplines ranging from ceramics to metallurgy to combustion chemistry.

In an old-school engine, the exact moment the spark plug fires is determined by a mechanical distributor, a device spinning in perfect synchronization with the camshaft. An electrical contact arm spins around once for every revolution of the camshaft (and thus once for every two revolutions of the crankshaft). Around the outside of the distributor cap is a set of contacts, one per cylinder, that delivers energy for the spark to each spark plug in sequence.

You might think that the perfect moment to fire the spark plug would be the moment the cylinder reaches top dead center; that way all of the explosion happens when the cylinder is traveling down, and the pressure is pushing the right way. But the explosion, while very fast, does take some time to get going after it's lit. So in fact it is desirable to send the spark just a bit *before* top dead center. How long before? That depends on how fast the flame will spread, and on how fast the cylinder is traveling. When the engine is running faster, you want the spark to go off sooner in the cycle, because by the time the explosion really gets going, the piston will have moved farther than when the engine is running slower. Clever mechanisms were invented to mechanically "advance" the timing based on engine speed, and of course in modern engines this is done by the computerized controller, which can take into account not only the speed of the engine, but also the fuel mixture and the temperature of the engine and incoming air.

The distributor cap is the high-voltage half of the distributor. It's tall to keep this high voltage away from the bottom half.

Spark plug wires connect the contacts to the spark plug in each cylinder.

▷ An old-school mechanical distributor handles two main functions: opening and closing an electrical contact once each time a spark is needed by one of the cylinders, and routing the spark to the correct cylinder each time. These two functions are handled by the bottom and top halves of the distributor, respectively.

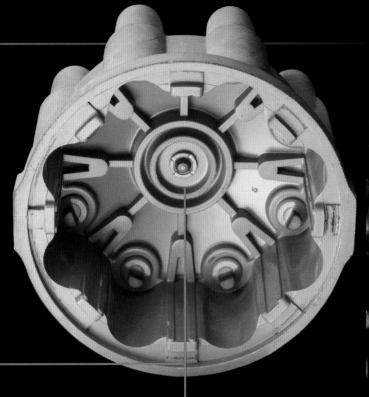

The center contact in the distributor cap touches the contact in the center of the wiper arm, carrying high-voltage spark energy from the spark coil.

△ An artistic rose made of spark plugs shows off the variety of styles available.

The wiper arm rotates once per revolution of the camshaft.

As the end of the wiper arm passes by each of the contacts in the distributor cap, it comes *close to* but does not actually touch the contact. The spark voltage is high enough to jump this small gap (just as it is high enough to jump the gap in the spark plug itself).

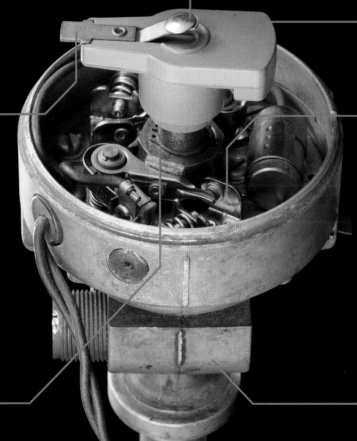

These contacts, called "points," open and close once each time a spark is called for, switching on and off a 12-volt DC "ignition coil" powered by the car's battery. Interestingly, the spark is created when the contact is *opened*, not when it's closed. To learn how the low voltage from the battery is turned into many thousands of volts, and why that happens when the connection is broken rather than when it's made, be sure to read the section on electricity, magnetism, induction, and transformers in the next chapter.

This cam rotates with the wiper arm: its octagonal shape alternately pushes on and releases the electrical contacts eight times per revolution, making and breaking the connection.

In here are mechanisms that advance the timing based on how fast the shaft is spinning, and how much vacuum is being created in the intake manifold (a measure of how hard the engine is working).

How to Break Your Friend's Nose
with Spark Advance

YOU KNOW THE argument about manual transmission versus automatic transmission? Some people like to micromanage every detail of their car, telling it exactly what gear to be in at every moment. Other, less controlling people prefer to give their car the freedom to make up its own mind about which gear would be best. It's more than just gears, though. Today, all cars have a fully automatic choke, but I remember riding in my parents' Sunbeam sports car in the 1970s, which had a manual choke that controlled the airflow into the engine and needed to be pulled out to start the car. People even older than me may remember the control shown here: a manual spark advance lever. It allowed you to fine-tune the combustion timing of your engine as it was running. Talk about neurotic micromanagement! It's like helicopter parenting for cars. But such things were useful, even necessary, when engines didn't have automatic mechanisms to keep themselves operating efficiently in a range of different conditions.

This fine old automobile, a 1912 Abbott-Detroit, has two unfamiliar (but common at the time) levers centered in its steering wheel: One controls the spark timing; the other controls the throttle, like the gas pedal in a modern car. Notice that the levers are fairly similar. When you're starting the car, you want to advance the throttle to give the engine plenty of fuel and air to work with. And you want to retard the spark so it goes off later than normal, when the piston is safely well beyond its top dead center position. You do not want to mix up the two levers and accidentally advance the spark instead of the throttle.

Before the invention of the electric starter motor you had to start your car with a crank on the front. Turning the crank rotates the crankshaft, pushing one cylinder after another through its compression stroke, until one of them fires successfully and the engine starts turning on its own. This can be quite difficult to do, because you have to overcome the resistance

Spark timing lever

Throttle lever

△ Steering wheels of the past didn't have buttons for cruise control and Bluetooth connections—they were busy taking care of more basic issues.

◁ Like all old cars, this one has stories—not all of them old.

△ This is a starting crank on a car with no electric starter. These things can be dangerous!

of the compression stroke. So you need a firm grip on the crank with your whole body behind it. If things go well and the engine starts up, a one-way ratchet in the crank lets the engine continue on in the forward direction without trying to make the crank spin with it. But that same ratchet necessarily does *not* let go of the crank if the engine tries to turn backward. If that happens, the handle of the crank will be rapidly and *forcefully* turned in the opposite direction.

So, as the story goes, Allen tried to start his car. He was at the controls while his friend Steve was doing the cranking. Allen confused the spark timing lever with the throttle. Steve broke his nose on the hood of the car when his whole body was violently jerked down by the force of a prematurely lit cylinder rotating the engine backward. And this, children, is why we mustn't confuse the throttle with the spark.

Plot twist: Steve of the broken nose was also the head of the chemistry department at the university I graduated from thirty-five years ago. He was my teacher for Chem 106, the first chemistry class I took in college—and that makes him partially responsible for my writing a book about the periodic table decades later, which ultimately led to this book with a paragraph about him breaking his nose. Neither Allen nor I knew there was any connection between me and Steve before Allen told me this story. That's how you can tell I live in a small town. Not *too* small, but, well, small enough.

Spark–Free Diesel Ignition

WE KNOW THAT AIR gets hot when it's compressed and that this can be a problem in gasoline engines because it can cause the fuel to ignite too soon and the engine to knock during the compression stroke. Diesel engines, on the other hand, avoid this problem by injecting the fuel into the cylinder only when it's time for it to start burning—just as the piston reaches the top during the compression stroke. At that moment the air is fully compressed and very hot, so the fuel lights immediately as it enters the cylinder. To put it another way, in a diesel engine the fuel is injected at the same point in the cycle as the spark plug would fire in a gasoline engine.

The challenge is that the fuel has to be injected at just about the most difficult moment in the cycle, when the air in the cylinder has been squeezed down to its maximum pressure. In order to withstand this high back-pressure, diesel engine fuel injectors are comprised of massive solid steel high-pressure pumps, driven by powerful cams. In addition, strong springs close up the port as soon as the fuel has been injected. That's important because, as high as the pressure is when the fuel is injected, it's about to get a lot higher when the fuel burns!

A gasoline engine injector is a flimsy bit of plastic compared to a diesel injector. It can be so small because it injects its dose of fuel when there is little or no pressure in the cylinder: it's like a squirt gun compared to a diesel injector's cannon shot.

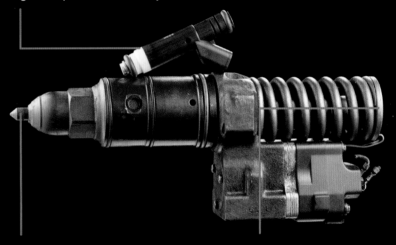

Diesel injector

A cam, similar to the ones that open the intake and exhaust valves, pushes the rod in against the large spring. Inside, this drives a piston forward, pushing fuel out through the small holes on the other side.

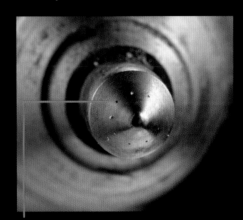

Diesel fuel enters the cylinder through this ring of tiny holes.

△ This is a slightly fancy injector that has a *solenoid*, which operates a low-pressure pre-pump that injects a measured amount of fuel into the main high-pressure pump cylinder in response to electronic signals from the engine management computer. The computer is optional: fully mechanical metering was used long before electronics.

As much research has gone into diesel fuel injectors as went into the design of spark plugs, but on the whole, they are much less finicky. The evidence for this is the fact that diesel engines can run on just about anything, from high-purity refined kerosene (hexane) to the waste sludge from oil refining, called bunker oil. There are no spark plug contacts to foul up with residue from impurities, and as long as the piston is sealed tightly to the cylinder wall, allowing pressure to build up during the compression stroke, it's almost impossible for the fuel not to light when it enters the superheated air. There is also no need for any electrical components, such as spark plugs, ignition coil, distributor, and so on. The entire engine can be made of nothing but steel.

About the only thing that can prevent a diesel engine from running is if it's too cold—so cold that the compressed air in the cylinder doesn't get hot enough to light the fuel. There are two common solutions: Start the engine up with a special blend of fuel, a lighter mix, that ignites more easily. Or heat up the engine using an electric block heater—or, if you're trapped in the Siberian wilderness with the wolves circling and no other way to get the engine started, by building a camp fire underneath it. By the way, this would be one of those times when the fact that the whole thing is made of steel is really useful. If you tried to do the same thing with a gasoline engine, you'd likely melt the distributor or ruin the spark plug wires.

▷ Bunker oil, also called marine diesel or #6 fuel oil, is a residual oil left over after the lighter, lower-boiling-point, more valuable fractions of crude oil have been boiled off in an oil refinery. It's cheap because it's almost a waste product of refining. The downside is that it burns slowly, so it can only be used in engines with very large, slow-moving cylinders, and it's so thick that tanks of it have to be preheated to make it liquid enough to pump into the engine. It also tends to contain a lot of impurities, which create pollution when it's burned.

▷ One of the most critical parts of an internal combustion engine is the sliding seal that sits between the piston and the cylinder. If there is the slightest gap, high-pressure, hot, corrosive combustion gases will leak through from inside the cylinder down into the crankcase. This will quickly erode the seal and destroy the engine. The solution is a set of "piston rings" made of very tough steel that fit into grooves in the piston. The spring tension in these rings keeps them pressed firmly against the inside of the piston, ensuring a tight seal even as the various components expand as the engine heats up. If you ever have a failure in the lubrication system—if the oil pump fails, or there's not enough oil in the engine—and these rings start sliding dry against the cylinder bore, your engine will be toast in a matter of seconds. The damage caused can be so complete that the engine might as well be thrown away.

This large piston has four separate piston rings to make sure nothing escapes.

The second and subsequent rings have slightly—by a few thousandths of an inch—smaller gaps, each according to its calculated operating temperature.

Fire Pistons

Compression heating, as used in diesel engines, is a very, very old way of starting a fire. "Fire pistons" made of bamboo or bone were invented in Southeast Asia around 2,500 years ago. A tiny bit of cotton or leaf "tinder" is put in the bottom of the piston, which is then pushed down as hard and as fast as possible. The heat generated by compression of the air lights the tinder on fire. The fire can then be transferred to a pile of leaves or dried grass, and from there to sticks and logs.

History records that Rudolf Diesel saw an Asian fire piston being demonstrated, and that this led to his invention of the diesel engine. Why didn't people make the leap from

starting fires to making engines two thousand years earlier? Because they didn't have the precision machining technology necessary to make a cylinder and piston with a perfect enough seal to allow such an engine to work for any length of time. The hard part of an internal combustion engine isn't the idea so much as it is the fact that these things work well only if they are very precisely made out of advanced forms of metal unavailable before the 1800s.

This demonstration fire piston is clear so you can see the ignition inside. You can buy brass versions for camping; they are more reliable than butane lighters, and easier than rubbing two sticks together.

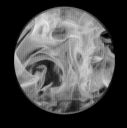

Piston from model airplane engine for comparison.

The size of each piston ring is *very* carefully controlled so that when it's inserted in the engine and at its normal operating temperature, this gap will close up precisely. If the ring is too long, it will bind against the cylinder wall. If it's too short, the gap will let hot gases through and destroy the engine. The topmost ring needs the largest gap, because it is closest to the heat and thus expands the most when the engine is running.

△ This is a piston ring. Get it? It's a ring with a piston, so it's a piston . . . ring. Just laugh and we can all move on to the exhaust stroke.

The Exhaust Stroke

AFTER ALL THE SOUND and fury of the combustion stroke, it's time to relax, exhale, and breathe out the waste gases to make room for a fresh breath of fuel and air in the following intake stroke. As the piston rises during the exhaust stroke, the exhaust valve opens and the contents of the cylinder are pushed out into the exhaust manifold (a set of pipes that routes the exhaust out and away from the engine, hopefully through a muffler so it's not too loud).

There's nothing particularly interesting about the exhaust stroke, but what's *in* the exhaust gases is of great concern. That's what gets dumped into the air that everyone has to breathe. Exhaust gases are made up of a complex mixture of chemicals, some—like water—entirely harmless, some of concern only when released on a global scale, and some quite toxic right here and now.

As soon as the piston reaches top dead center on the exhaust stroke we are back to where we started, at the beginning of the next intake stroke. For an engine running at a highway speed of 3,000 rpm, the whole thing, all four strokes, takes $1/25$ of one second, with just $1/100$ of a second allotted to each of the four strokes!

▷ In a perfect world, fuel would contain nothing but hydrocarbons (molecules containing only carbon and hydrogen) and the only exhaust products would be CO_2 (carbon dioxide) and water. Sadly, this isn't the case. All fuel contains some impurities, particularly sulfur, and combustion is never perfect, leaving undesirable nasties in the exhaust stream.

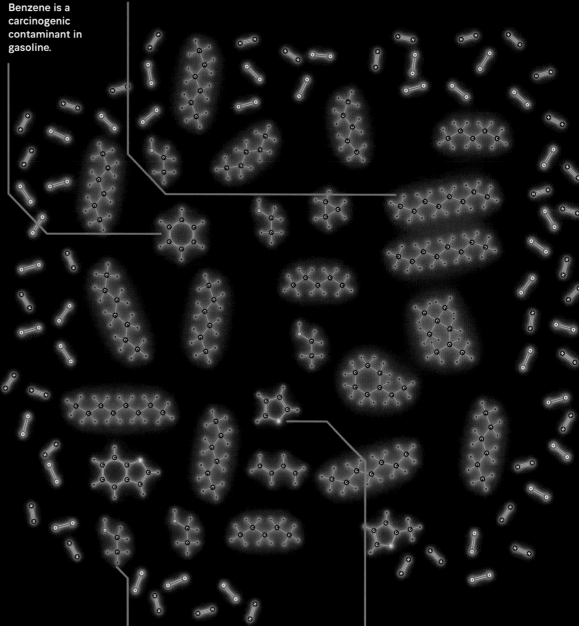

Benzene is a carcinogenic contaminant in gasoline.

1 Hydrocarbons
These hydrocarbons—straight, branched, and cyclic—are what fuel is supposed to be made of.

▽ Air is about 20 percent oxygen (O_2), which combines with fuel, and 80 percent nitrogen (N_2), which is just along for the ride.

2 Ethanol
Some fuels also have ethanol, which contains oxygen along with carbon and hydrogen. This is fine.

3 Sulfur Compounds
Sulfur embedded in cyclic aromatic molecules is a common contaminant in fuels, particularly diesel fuel. This is bad, because it leads to sulfur dioxide in the exhaust.

▷ The inside of a catalytic converter

4 Carbon monoxide

Carbon monoxide is created in larger amounts when there isn't enough air for the amount of fuel in the cylinder. It's extremely toxic to humans and other animals because it binds with red blood cells and prevents them from transporting oxygen. If you're locked in a closed room with a running gasoline engine, it's the carbon monoxide that will kill you first.

7 NO & N₂O

Nitrogen oxides are created in larger amounts when combustion occurs at higher temperatures. They interact with sunlight to create the smog that used to blanket cities, like Los Angeles, before stricter laws forced car makers to redesign engines with pollution control devices.

5 Sulfur oxides

Sulfur in the fuel turns into sulfur dioxide and sulfur trioxide. These are bad things to put into the air. Combined with water, they form sulfuric acid, which leads to serious ecological damage and breathing problems. If you don't have asthma and want to know what it feels like, try breathing around some burning sulfur.

6 CO₂ & H₂O

Carbon dioxide and water are the main exhaust gases. In moderation, neither is much of a problem.

Unburned fuel molecules in car exhaust are a major source of air pollution.

THE MOST IMPORTANT pollution control device in a modern car is the catalytic converter. Typically these devices contain tiny beads of platinum, rhodium, and/or palladium (embedded in a ceramic grid) that have the peculiar property of causing leftover partially burned fuel to finish combining with oxygen. This is a very effective way of reducing the amount of unburned hydrocarbons that make it out of the tailpipe. The converter can also add an extra oxygen atom to carbon monoxide (CO), turning it into the much less harmful carbon dioxide (CO_2), and it can strip the oxygen from NO and N_2O molecules, turning them into completely harmless N_2 and O_2 molecules. If you're in a major city anywhere in the world and the air is clean and clear, that is largely thanks to laws requiring catalytic converters in cars and trucks.

The 411 on CO₂

For decades people worried about cars spewing out anything that wasn't CO_2 or water. Getting everything fully burned up into those two compounds was the definition of success, because everyone assumed both were harmless. And in a local sense they are: Water is completely harmless, and CO_2 is something we willingly drink up and burp out every time we have a carbonated soda. It wasn't until the 1970s and 1980s that we started to realize that, in the long run, emitting a *lot* of CO_2 from many sources could have an entirely unhappy effect, that of disrupting the global climate. We're currently living in that long run, and that's one of the many reasons internal combustion engines are not long for this planet.

▽ **Both valves closed**

▽ **Exhaust valve open**

▽ **Piston traveling up, pushing out exhaust gases**

▽ **Exhaust valve closed again**

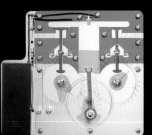
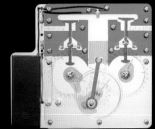
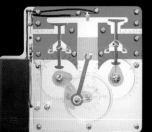
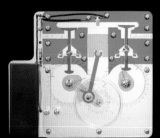
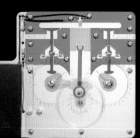

Two-Stroke Engines

THE GREAT MAJORITY of engines you encounter in the world, including essentially all car engines, are of the four-stroke variety. But there is an alternative. Two-stroke engines can be much less complicated than four-stroke engines. *Much* less. They don't need any timing gears or belts, camshafts, or complex valves. And because they fire their spark plug every time the piston reaches top dead center (versus only every other time in a four-stroke engine) they can in principle generate twice as much output power for a given cylinder size.

Unfortunately, the price you pay for this simplicity is lower fuel efficiency and much dirtier operation, with large amounts of soot and unburned fuel blown out the exhaust. Two-stroke engines are common on small power equipment (push lawn mowers, hedge trimmers, etc.) but are not used on modern cars or even on better-quality lawn mowers. According to a study by automotive research firm Edmunds, "Hydrocarbon emissions from a half hour of yard work with the two-stroke leaf blower are about the same as a 3,900-mile drive from Texas to Alaska in a Ford F-150 pickup truck."

Despite being mechanically simpler, two-stroke engines are actually a good bit more difficult to understand.

To help us understand the logic of how these engines work, I have made this diagram that simplifies the operation down to its bare essence. Simple two-stroke engines are surprisingly subtle in their operation, because of the way they use the bottom side of the moving piston as a pump. You have to follow the flow of fuel, air, and exhaust gases not only through the interior of the cylinder, but also through the crankcase. You will notice a complete lack of valve hardware: no cams, no timing gears, and so forth. In fact, it almost looks like there are no valves at all. The "valves" are actually just holes in the sides of the cylinder wall, called ports, which are covered and uncovered by the piston itself as it moves up and down.

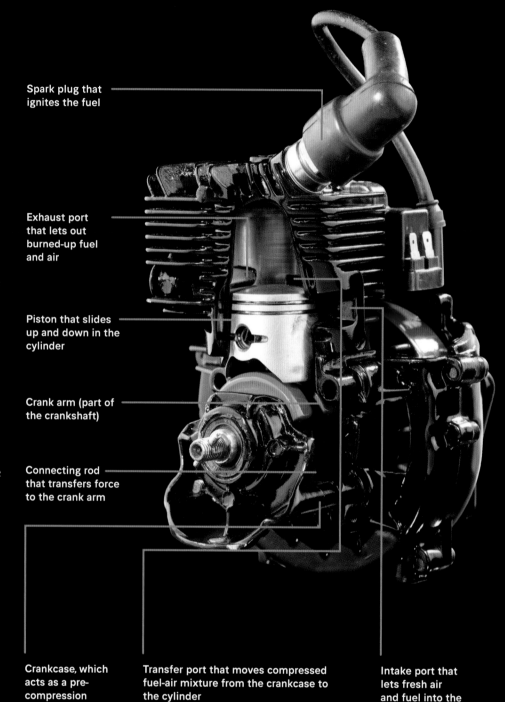

Spark plug that ignites the fuel

Exhaust port that lets out burned-up fuel and air

Piston that slides up and down in the cylinder

Crank arm (part of the crankshaft)

Connecting rod that transfers force to the crank arm

Crankcase, which acts as a pre-compression chamber

Transfer port that moves compressed fuel-air mixture from the crankcase to the cylinder

Intake port that lets fresh air and fuel into the crankcase (*not* into the cylinder!)

△ This is a real two-stroke, single-cylinder engine with sections of its body cut away to show the working parts inside. Once again, we'll compare the real two-stroke engine to the acrylic model and try to imagine being able to see inside the solid metal version as if it were clear.

△ We start with the most exciting moment: The spark plug has just ignited a charge of compressed fuel and air in the tiny space above the cylinder. The main action at this point is the pressure of the ignited fuel pushing down on the piston, delivering power to the crankshaft. While the explosion is going on upstairs, there's something else happening down below. With the piston at the top position, a gap has opened up allowing outside air to enter the crankcase *below* the piston. At the point that this gap opens up, there is negative pressure (a partial vacuum) inside the crankcase, so air rushes in, pulling fuel from the carburetor in with it.

△ After the cylinder has moved down a bit, the intake valve at the bottom is closed off, and the crankcase is sealed off from the outside air. (The crankcase in a two-stroke engine must be completely airtight or it won't work.) As the cylinder continues to travel down, the fuel-air mixture within the crankcase is compressed—not in the cylinder, but *below* the cylinder.

△ With the cylinder nearing the bottom of its stroke, it exposes the exhaust port on the cylinder wall and exhaust gases start flowing out. At this point the cylinder above is open to the outside, and in the crankcase below, the fuel-air mixture is pressurized from the downward movement of the piston.

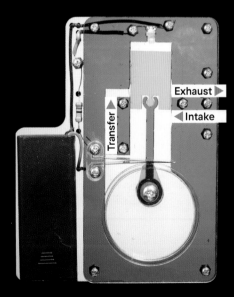

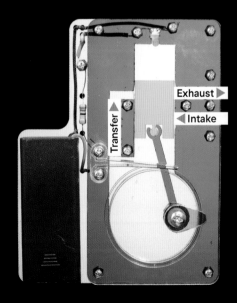

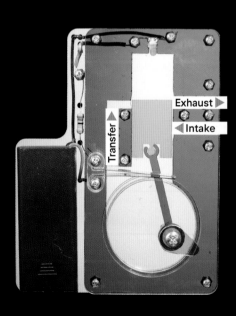

△ Just a bit farther down and the transfer port is exposed. This opens up a connection between the crankcase and the cylinder. The compressed fuel-air mixture in the crankcase now rushes up through the transfer port and into the cylinder. But watch out! The exhaust valve is still open. It has to be, because we are relying on the inrushing fresh fuel-air mix to push out the rest of the exhaust gases. Inevitably, some unburned fuel will make it out the exhaust port, leading to one source of pollution: unburned fuel in the exhaust.

△ A bit later the piston is now heading back up. The transfer port is closed. The exhaust port stays open a tiny bit longer, and if the exhaust system is designed correctly, during this brief moment back-pressure from the exhaust will actually push most of the unburned fuel back into the cylinder just before the exhaust port closes.

△ As the piston travels upward, two things happen. First, the volume *above* the piston gets smaller, compressing the fuel-air mixture inside (just like in a four-stroke engine during its compression stroke). Second, the volume *below* the piston gets larger, creating the vacuum in the crankcase that we saw earlier.

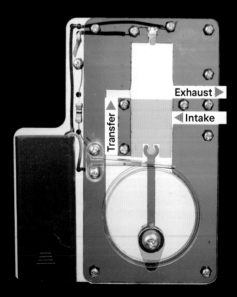
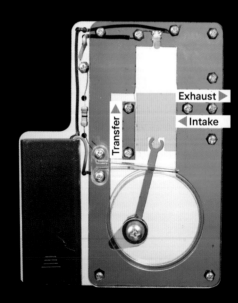
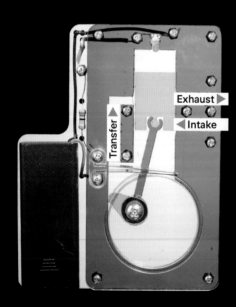

THE FOUR-STROKE ENGINE separates each of the four tasks necessary to make a complete cycle: (1) draw in fresh fuel-air mix, (2) compress it, (3) ignite and expand it, then (4) flush out the exhaust. A two-stroke engine cleverly does part of the compression work on the bottom side of the cylinder at the same time the ignition/expansion part is happening on the top side.

Then it does both the intake and exhaust phases more or less instantaneously while the cylinder is at the bottom, eliminating two whole strokes from the cycle. It's this combination that results in most of the challenges associated with two-stroke engines. There's just no way to completely separate the exhaust from the incoming fuel, so inevitably some unburned fuel escapes.

Using the crankcase to do pre-compression is what lets you avoid a separate intake stroke, but this, too, is problematic. Piston rod bearings and crankshaft bearings *must* be continuously lubricated with oil. They are under tremendous pressure and operating at high speed. Without oil, they would burn up in minutes. Gasoline, unfortunately, dissolves and volatilizes oil. So, when the fuel-air mixture is being compressed in the crankcase, it's also dissolving away the oil and flushing it into the cylinder, where it is burned up, creating more pollution.

To keep the crankcase oiled even as the gasoline is trying to de-oil it, you have to mix oil in with the gasoline you're using. This is why two-cycle engines can't burn ordinary gasoline. The fuel being drawn in on each stroke must supply enough oil to keep the crankshaft

lubricated. This oil is formulated to burn as cleanly as possible, but it's never possible to make a heavy oil burn as cleanly as the much lighter, smaller molecules in gasoline.

△ Just before the piston reaches the top and the spark plug is fired, the intake port opens up, allowing air and fuel to rush in to fill the vacuum in the crankcase. We are back to the start of the cycle, and now you know how we got there! The intake port will remain open until the cylinder has moved a little way down, closing it off again and beginning the compression of the fuel-air mix in the crankcase.

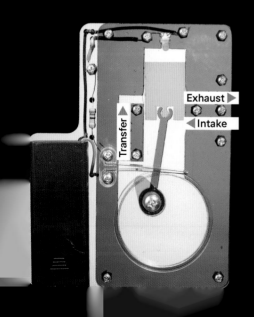

△ How simple can a two-stroke engine be? Here are *all* the parts of a small model aircraft engine laid out for you to see. Including all the screws, washers, and so on, there are only about twenty parts in the whole thing! That's actually fewer parts than in my acrylic model, mainly because my model is held together with a bunch of separate screws whereas the real thing has a few complicated three-dimensional parts. This engine could even be made with several fewer parts: two parts form a flapper-style intake valve that could have been eliminated by using a sliding port valve, with zero parts, instead.

▷ Surprisingly, there is one place you will find very large, relatively efficient two-stroke engines, and it's at the absolute opposite end of the spectrum from the typical lawn mower or model airplane engine: massive, low-rpm marine diesel engines. This is possible because when the cylinders are huge and operating very slowly (only about a hundred times per minute) it's possible to surround them with extra machinery that carefully manages the delivery of fuel and air to optimize the performance of each cycle. In this engine the crankshaft is completely isolated from the cylinder above, so it can be lubricated in the conventional way. Compression is done with an entirely separate turbocharging mechanism that injects fuel and air directly into the cylinder.

A Visit to the Showroom of Allen and Nancy Strong

INTERNAL COMBUSTION engines exist for one reason beyond all others: cars. This was their first killer application and remains, by far, their most important job to this day. People *love* their cars. For some people, the grease monkeys, it's the engines inside that are the real attraction. For others it's the style, the speed, or the luxury. For true car lovers it's the whole package, and rarely do you see a package as whole as the one I stumbled on just a few miles from my house.

△ Your first impression on entering this shed-in-name-only is that you have entered a time warp. Just about everything in here would be perfectly at home in a luxury car dealership of the 1920s or 1930s, including the owner, Allen, seen at the far left in his natural habitat. At that time the Ford Model T had been available for many years (since 1908) and was a popular and affordable car—but you won't find one here. These are the high-end luxury cars, race cars, and flashy sports cars of their age.

This car was owned by the person who built the Indianapolis Motor Speedway, the racetrack for the Indianapolis 500. It was the pace car for the first running of that famous race.

This door leads to a *second* room full of cars.

Only two examples of this type of car are known to still exist; the other is in the collection of Jay Leno, the talk show host turned extreme car collector.

△ When you see a door like this, expectations are not high. But the first time I walked through this particular one I was absolutely blown away, and I hope I can show you in these pages a bit of the magic of this place. Behold the inside of Allen and Nancy Strong's garden shed.

The second room is no less spectacular than the first.

This is the only car of its kind known to exist in the world.

1913 Cadillac

THIS 1913 CADILLAC has just about the most beautiful engine I've ever seen. It's distinguished by having won the Dewar Trophy (an important automotive competition in the United Kingdom) in 1908 for . . . having completely interchangeable parts. This was a significant innovation in an age when most cars were still craft-built one at a time, with parts carefully fitted to each other, not to a uniform standard. They took three cars at random from the manufacturer's inventory, disassembled them thoroughly, scrambled the parts, and then reassembled the heap of parts into three entirely identical—but entirely different—cars that were able to complete a test run. This is the only extant 1913 Cadillac Roadster.

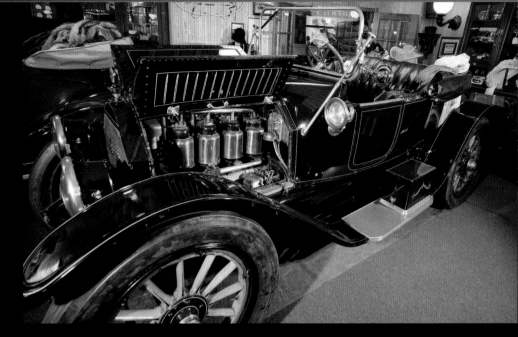

△ Just look at this beauty! Nearly all automobile engines are water-cooled. They have channels surrounding the cylinders filled with water that is circulated by a pump through the engine and then through a radiator that cools it. Most engines embed these channels invisibly in the cast-iron cylinder block, but not this one! Its cooling water flows through gorgeous copper cans, connected by soldered copper pipes (not unlike copper house plumbing pipes).

△ One of the many hassles of owning a car in the early days was that rubber tire technology was not yet up to the challenge. If you took your car out for a drive to the next town, you would be nearly guaranteed to have a flat somewhere along the way. Most cars came equipped with a hand pump, similar to today's bicycle pump. But not this Cadillac! As one of the high-end features of the day, it had a little air compressor built right into the engine. A sliding gear engaged it so you could pump up your flat tire in Cadillac luxury.

△ Here's a further nod to the fact that these early cars were not just "turn-the-key-and-drive" machines: a small oil can holder that ensures you are always ready to oil that thing that always needs oiling.

1915 Packard

THIS CAR HAS a deep and significant history in the early days of the American car industry. It was custom-built for automotive entrepreneur Carl Graham Fisher, to serve as both his personal car and his sports car for auto racing. Carl was someone who thought big. He felt that the American car industry was being held back by the lack of a good test track for proving the endurance of cars, so he built the Indianapolis Motor Speedway and challenged people to race on it. They're still doing that today.

Another thing that Carl felt was holding back the American car industry was the lack of good cross-country roads that would let people drive from coast to coast without getting stuck in the mud somewhere in Kansas. So he campaigned for years to raise the funds to build the Lincoln Highway, one of the first cross-country paved roads.

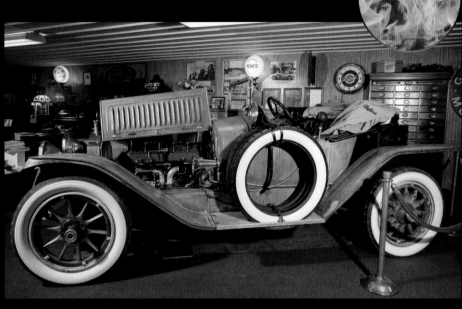

△ This 1915 Packard was custom-built for Carl Graham Fisher.

▷ Here is a picture of this car—not a car like it but this *actual* car—acting as the pace car for the 1915 running of the Indianapolis 500 and driven by Carl Fisher himself.

△ Here's a picture of Carl Fisher with this car—not a car like it but this *very same* car—in Elkhart, Indiana, in 1915. He used it to survey the route of the Lincoln Highway after the road was dedicated and work got underway to actually build it.

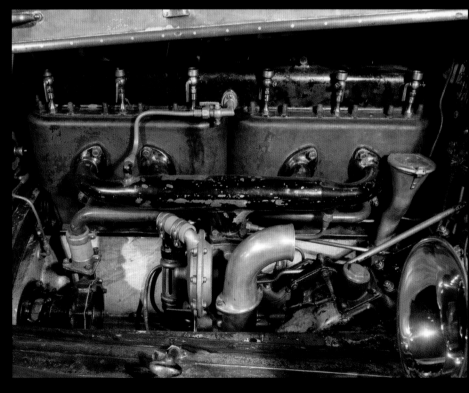

△ Carl's Packard has a six-cylinder engine with all the cylinders in a row (a "straight six"). Notice that there are two separate groups of three cylinders, which was common in early engines. This is because the metallurgy and casting technology of the time would have been hard-pressed to make a successful casting as large as the whole engine in one go. It also made servicing the engine easier: you only had to take apart half as much of the engine to get at a given cylinder. To see the modern configuration, we need to move on to the next car.

1927 Packard

EIGHT YEARS AFTER Carl Fisher's iconic 1915 Packard, this 1927 model Packard shows how quickly automotive technology was advancing during that era.

1930 Cadillac

HERE'S A GORGEOUS 1930 Cadillac convertible with a fun feature: a second windshield between the two rows of seats, which can be lowered when the roof is up. And there's a surprise hiding behind it.

△ The 1915 Packard had two cylinder blocks that included everything from top to bottom for three cylinders each. The 1927 model has eight cylinders, but they are all in one block. Instead of splitting the block vertically into separate groups of cylinders, they split the block horizontally into a cylinder head on top, cylinder block in the center, and crankcase on the bottom. The advantage of this configuration is that many of the parts likely to break—the valves, cams, timing gears, rocker arms, and so on are immediately accessible by removing the cap on the top of the engine. And because the crankcase isn't subject to high-temperature exploding gasoline, it can be made of much lighter—but lower-melting—aluminum. This is how nearly all modern engines are set up, with three major castings from top to bottom.

△ This car takes backseat driving to a new level, with an actual backseat dashboard! A speedometer lets you yell at the driver for going too fast or too slow, backed up with real data. Actually, the speedometer and chronometer (an accurate clock) could be used by navigators during surveying or cross-country races.

1934 Packard

THE BLUE OF this 1934 Packard is just gorgeous!

1935 Auburn Supercharged Speedster

THIS 1935 AUBURN Supercharged Speedster was, of course, super-charged, and every one was tested at over 100 mph before being offered for sale. Like a Lamborghini today, it wasn't really a serious car, it was a toy for playing out on the road. Not saying there's anything wrong with that.

△ This engine was obviously designed to be looked at. Like the hot rods of today, and of every decade stretching back to the first fast cars, it's not enough to just go fast, you have to *look* like you go fast.

△ By 1934 engines were starting to look positively modern, if modern engines were made this beautifully! Just look at the intake and exhaust manifolds (air channels) on top: that's just showing off.

△ People have been picking particular aspects of their products to use as marketing hooks for as long as there have been products needing marketing. Chevrolet has their "Hemi" engine, which refers to the hemispherical shape of the top of the combustion chamber. DeWalt decided to put the word "Brushless" on the side of their tools in letters *larger* than their own logo, referring to the type of electric motor inside. And Auburn proudly labeled this car as SUPER-CHARGED, to be sure prospective customers wouldn't miss how cool that is.

BEFORE WE LEAVE the Strongs' shed, we must appreciate all the things in here that are not, strictly speaking, cars. Nancy has filled the space with period-appropriate costumes, furs, and hats. They, along with the cars of course, really give the place a magical time-warping quality. This is what a lifelong of hard work and joyful collecting looks like.

▽ This is definitely the biggest music box I've ever seen.

△ Steam engine model

▽ So many hats!

I LEFT THIS little air filter from the Strongs' collection until the end because it's most interesting not for how it's used in a car, but for how it explains the function of the face masks that are so much a part of life as I write these words during the COVID-19 pandemic.

Engines do not like dust in the air they suck in to combine with the gasoline they burn. Tiny particles of grit, such as you might kick up driving on a dirt road, can scratch the interior surfaces of the cylinders, melt and stick to surfaces in the exhaust system, and generally cause problems. An air filter is a crucial component.

These days they are typically made of a nonwoven pleated material, a lot like furnace and vacuum cleaner filters. But the older ones work in a completely different way. Instead of using a filter with very small holes, they use a mesh of steel wool with gaps between the wires that are *much* too big to trap the particles in question.

How does a coarse wire mesh stop particles that easily fit through the holes? The secret is that there are many layers of wire mesh, and the wires are all coated with a thin layer of motor oil, which makes them sticky. A particle might make it through one or two layers, but the chances of making it all the way through the filter without hitting and sticking to a wire somewhere along the way are very small.

Face masks made to filter out virus particles have an even more challenging problem. Because the virus is so small, any filter with holes small enough to block it would also

be impossible to breathe through. So protective face masks work the same way the old engine air filter works: The fibers they're made from are sticky. They aren't coated with engine oil; instead, they are made of an "electret" material. Electrets are the electronic equivalent of permanent magnets. Instead of having a permanent, embedded magnetic field, they have a permanent electrostatic field built in.

Inside the electret fiber are countless individual molecules that have an imbalance of electric charge from one side to the other; that is, one end is more positively charged than the other. When the material is in a molten state, these molecules can rotate, but when it's solid, they are locked into position. To make an electret fiber, the material is heated, drawn into a fiber, and then allowed to cool while sandwiched between two metal plates with a very high electric potential between them. The molecules align to the electric field and are then locked in that orientation when the material solidifies. This creates a fiber that is forever locked into having a net positive charge on one side and a net negative charge on the other. It's like a dress with static cling that you just *can't* get rid of.

The permanent electric field attracts and traps tiny virus particles just like oil in the old air filters did, except on a scale hundreds of times smaller. It's remarkable how well these things work, but you do have to watch out because once the fibers are coated the electric charge is neutralized, and you can't heat them too much or the molecules will reorient, destroying the electrostatic field.

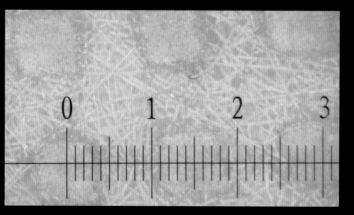

△ The scale in this close-up of an N95 face mask is in millimeters (so one of the small divisions is 0.1 mm). You can see that the gaps between fibers are about the size of one of these small divisions. This is a problem because a typical virus particle is only about 100 nanometers (billionths of a meter) across, which is *ten thousand times* smaller than the gaps. And yet the mask works quite well because the fibers attract and trap the virus particles electrostatically.

Models of Power

IF YOU DON'T HAVE the budget or the building to collect antique cars, there are lots of beautiful models of internal combustion engines you can get instead. Since the invention of engines, models have been used to help people figure them out. Models come in three basic variations: transparent, cutaway, and operational.

Transparent models are the least realistic, but the most helpful for learning purposes. As part of writing the book I decided to design my own transparent model, mainly as a way of learning firsthand how engines fit together. My first design is a three-cylinder, dual overhead cam model. Large timing gears connect the crankshaft to the two camshafts (one for the exhaust valves and one for the intake valves). A brass contact on one of the timing gears simulates the distributor: it touches screw heads to "fire" the LED simulated spark plugs.

After I finished the first model, I realized that it was really hard to see the three cylinders working together. I wanted to make a model where I could see the motion of *all* the parts—three cylinders, three spark plugs, six valves, the crankshaft, and the camshaft—from a *single* viewpoint. It took me weeks to finally figure out how to make a completely flat design, but I'm very happy with the result. The proportions aren't very realistic, but watching it turn lets me feel the rhythms of the engine in a way I had not experienced before.

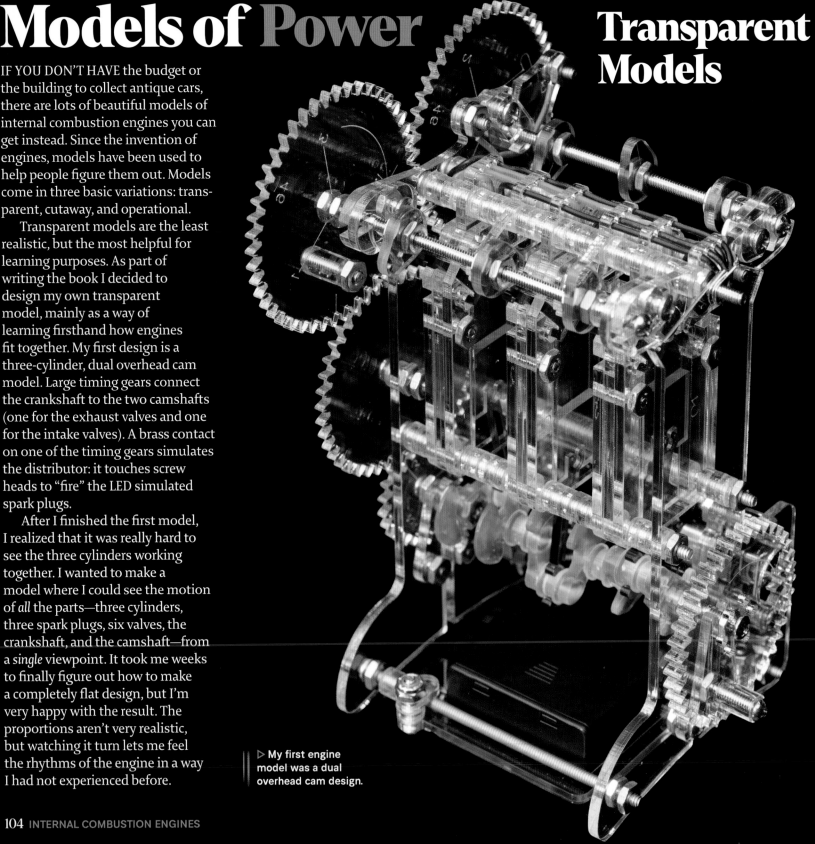

▷ My first engine model was a dual overhead cam design.

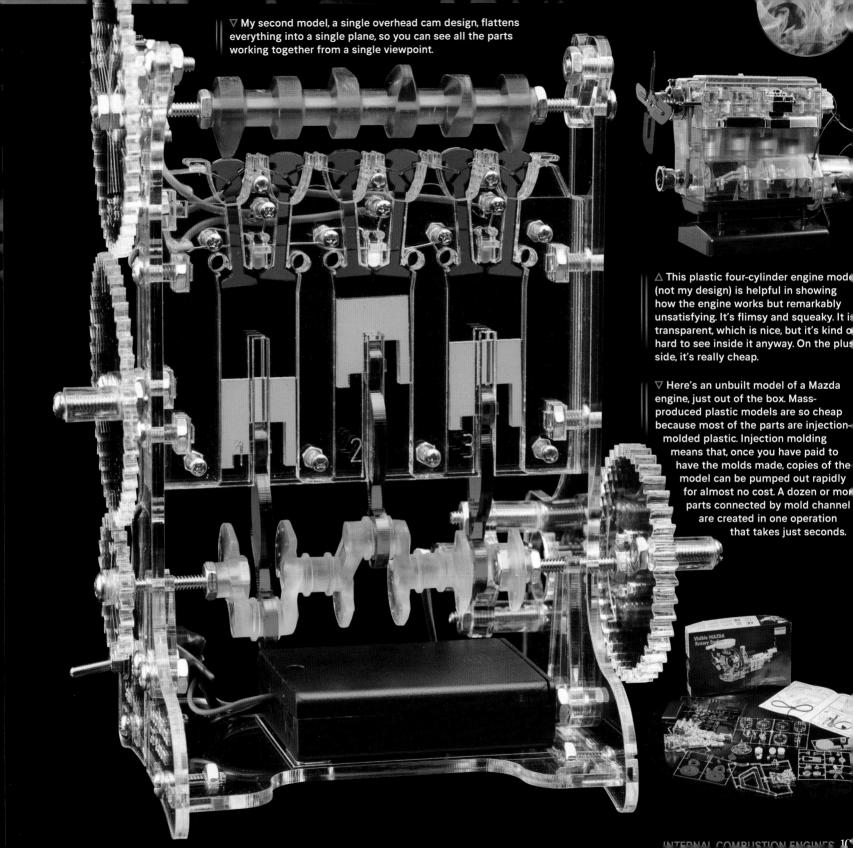

▽ My second model, a single overhead cam design, flattens everything into a single plane, so you can see all the parts working together from a single viewpoint.

△ This plastic four-cylinder engine model (not my design) is helpful in showing how the engine works but remarkably unsatisfying. It's flimsy and squeaky. It is transparent, which is nice, but it's kind of hard to see inside it anyway. On the plus side, it's really cheap.

▽ Here's an unbuilt model of a Mazda engine, just out of the box. Mass-produced plastic models are so cheap because most of the parts are injection-molded plastic. Injection molding means that, once you have paid to have the molds made, copies of the model can be pumped out rapidly for almost no cost. A dozen or more parts connected by mold channels are created in one operation that takes just seconds.

Cutaway Models

IF YOU WANT A more satisfying all-metal engine model, but you still want to be able to see inside, you need a cutaway. At the beginning of this chapter we saw some examples of real engines (small ones) that had been cut open to show their insides. For larger engines, there are models specifically designed as educational cutaways.

These lovely die-cast metal engine models are made by a company in China and retail for several hundred dollars. They come in a million pieces, with detailed instructions for putting them together. This is an important aspect of the value of a model: putting it together teaches you as much or more than watching it run when it's done.

▷ This is a cutaway four-cylinder, sixteen-valve metal engine model.

Every four-stroke engine in this chapter uses what are called "poppet" valves, the kind that are pushed down to create an opening in the cylinder. They are the universal type of valve used in nearly every engine. But there was a precursor to the poppet valve that was widely used for a time, known as a sleeve valve. These are mechanically simpler and completely avoid the use of valve springs. Two concentric sleeves around the piston slide up and down, opening and closing the intake and exhaust ports on the side of the cylinder. Two sleeves, moving at different times, are needed to create the correct sequence of openings and closings; a port is open only if the slots in both sleeves align at the same time.

This is similar to the way the ports work in a two-stroke engine, but in that case it works out that you don't need any sleeves, because the movement of the piston itself can cover and uncover strategically placed ports at just the right time. Two-stroke engines use ports to this day, but this more complicated double-sleeve design proved too unreliable.

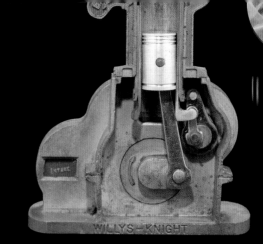

◁ **Intake stroke** intake port is open, exhaust port is closed.

◁ **Compression stroke:** both ports are closed.

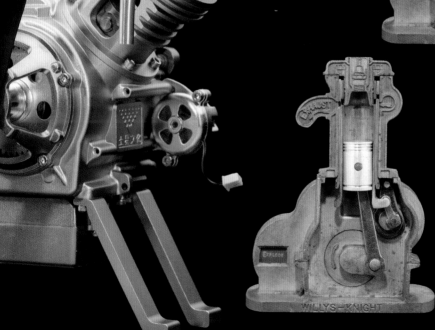

△ **This is a cutaway two-cylinder radial engine.**

△ **Combustion (called explosion in this model) stroke:** both ports still closed.

△ **Exhaust stroke:** intake valve is closed, exhaust valve is open.

Working Models

FULLY OPERATIONAL steam engine and Stirling engine models, as we saw in the first chapter, are very common, and pleasant to be around. They tend to be fairly quiet, and they can run on clean-burning alcohol lamps that give off almost no smoke. This is not the case with model internal combustion engines. These things are necessarily noisy, and unavoidably give off smelly and potentially dangerous exhaust, no matter how beautifully they are made.

Small working internal combustion engines tend to be found in places where they are actually useful—model airplanes or garden tools, for instance—but there are some examples designed specifically to be educational models.

These two weights spin with the flywheel. When the flywheel is going fast, centrifugal force pulls the weights apart against the springs that are holding them together. This is in effect a way of measuring the speed of the flywheel: the faster it's going, the farther apart the weights are.

I tried not to touch it before we took pictures, but it was just so seductive that I couldn't help myself.

▷ This beautiful working model of a hit-and-miss engine is brand-new, made in China, and available for about $300. It is every bit as well made as the best models of the past, and includes a lovely array of fully functional components: it will actually run.

The muffler reduces noise.

The fuel tank is made of brass.

Cooling water circulates in the space between the inner and outer wall of the cylinder housing.

A cam on the crankshaft operates this pump, which circulates the cooling water through the radiator and cylinder jacket.

The radiator cools the water circulating around the cylinder. (This is a proper water-cooled engine.)

▽ On the other side of the flywheel we see how the speed measurement is used. When the flywheel is stationary or spinning slowly, the grooved pulley is toward the left. When the engine is going fast, the spreading weights shift the pulley toward the right. This moves a lever back and forth, such that it either does or does not catch the exhaust valve lifter.

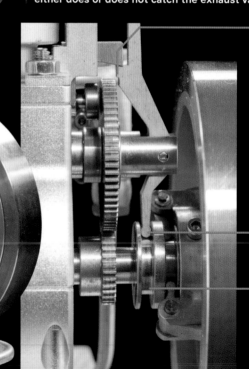

In this position, with the engine running fast, the bent lever catches the lifter and prevents it from following the cam back down.

This lever moves with the grooved pulley.

This grooved pulley moves to the right when the flywheel speeds up.

▷ When the engine is going slow, the exhaust valve is opened by the cam only when it normally should be, during the exhaust stroke. (Notice that in this position the grooved pulley is to the left, and the bent rod is not catching the lifter.) But when the engine is going fast, the bent lever catches the lifter (as in the picture) and doesn't let it back down, keeping the exhaust valve open during the following intake stroke. It's much easier for air to flow in through the exhaust valve than through the intake valve, so instead of filling with a mixture of gas and air, the cylinder fills with plain air, and no explosion occurs the next time the spark plug fires. As the engine slows down, the bent lever eventually moves out, and the exhaust valve closes. Then gasoline is sucked in, and the engine gets another kick of power. This is the origin of the name "hit and miss engine." It doesn't fire, or hit, every cycle. Under light load, most of the cycles "miss."

The exhaust valve normally stays open only during the exhaust stroke but is kept open longer when the engine is running fast.

The rocker arm changes the upward motion of the push rod into the downward motion of the exhaust valve.

The spark plug fires on every cycle, whether there is any fuel to ignite or not. This is called a "wasted spark" design. Some engines include a mechanism that extends the spark plug's life by firing it only when needed.

The intake valve is never mechanically opened, as in most engines. Instead, vacuum suction created by the retreating piston during the intake stroke pulls the valve down enough to slurp in some air and fuel mixture, as long as the exhaust valve isn't also open.

Fuel is mixed with incoming air in this carburetor.

This throttle valve controls how much fuel gets sucked into the engine on each intake stroke. This doesn't control the speed of the engine; the spinning weights do that. Instead, the amount of fuel controls what fraction of the cycles of the engine need to fire in order to keep that speed, depending on the load placed on the engine. If the engine is under light load and the throttle is wide open, the engine may only need to fire once every ten cycles. As the load increases, the engine will slow down more quickly after each firing of the cylinder, and a larger and larger fraction of the cycles will fire. To keep a relatively steady speed, you want to adjust the throttle so that most, but not all, of the cycles fire. (If the engine is firing all the time, that means it isn't able to get up to its intended speed.)

This cam lifts the lifter and push rod—and thus opens the exhaust valve—regularly for each exhaust stroke. But when the engine is running fast, the lifter is held up away from the cam, and the exhaust valve stays open during one or more whole cycles.

For the Love of a Machine

△ This is one of the largest shows in the country and is held every two years. It's impossible to overstate the sheer number and variety of old machines that come rolling by. You may notice that nearly everyone is in a golf cart or motorized buggy of some sort. That's because the show is so vast that at a brisk walk it takes a good half hour to get from one end to the other.

I RECENTLY SPENT a sunburned afternoon out on the airfield of an abandoned military base not far from my house, watching a parade of old tractors go by for well over an hour. One after another the great beasts lumbered by, most powered by gasoline or diesel fuel, but an honored few belching steam and coal smoke from their boilers and steam pistons. Surrounding me on all sides were thousands and thousands of people, all watching and commenting on the variety of engine types, and discussing whose uncle had one of those back in the day. The only child I saw the whole time was one ridiculously proud kid driving a real monster of a machine down the parade route.

There are dozens of these events every year in the US, attracting tens of thousands of visitors. What draws people to them? It is the love of the machines of their youth. The attendees are old people, by and large, remembering the days when these machines worked for them, their parents, and grandparents. The family tractor is more than a tool. Like the family horse that came before it, the tractor is the foundation of a farm family's livelihood and an investment perhaps worth more than the house they live in. This is true even today: a modern piece of farm machinery, say a combine harvester, can cost over half a million dollars.

△ People looking at dirt...that has recently been turned over by one of the many steam and gasoline tractors showing off their pulling power in the organized plowing demonstrations.

This whole tractor is run off a single high-pressure, dual-acting cylinder.

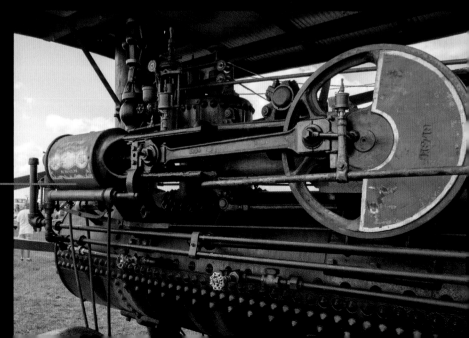

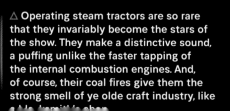

△ Operating steam tractors are so rare that they invariably become the stars of the show. They make a distinctive sound, a puffing unlike the faster tapping of the internal combustion engines. And, of course, their coal fires give them the strong smell of ye olde craft industry, like

△ Gasoline engine tractors make a beautiful sound as well. This old Avery 40-80 sounds like the rhythm section of a cool jazz band. It has a four-cylinder horizontally opposed engine. The exposed valves on the front set the base rhythm, syncopated by the softer valve sounds from the back side, and the duller thuds of the ignition strokes.

▷ Tractors are not like cars: They don't have thin sheet metal shells that rust out and turn to worthless scrap in a decade or two. Sure, some rubber parts dry out, seals need to be replaced, the battery wears out, and the oil needs changing. But the bones of the machine are strong, built to work hard for generations. A good tractor can easily outlast the working life of its first owner, even if it's given only minimal care and love. Tractor restoration companies take care and love to the next level, turning a tired but sturdy old machine into a gleaming beauty that lives up to the next generation's rose-

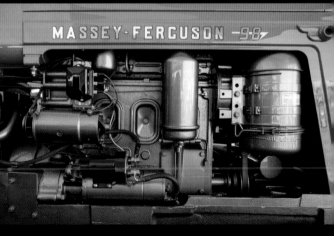

△ No working tractor was ever this perfect, but in the eyes of the child watching his parent mount the beast, so high up, it was perfect in every way. Here they have brought the machine back to match not what it was, but how it lives in the memories of those who love it. They have made it worthy of the life, forever

▽ In contrast to those lovingly restored models, this is my tractor. It was made ten years before I was born, and I bought it when it was already decades old. I have not been a good owner to it: I've left it outdoors, I've failed to maintain it sometimes for years. But it starts, and it runs, and when it needs fixing it's never anything fundamental. I feel bad about how poorly I treat my tractor, but realistically it's just not important enough to me to put in the time and money it would take to make it look like new again. It wasn't my parents' tractor. (They were both university professors, mathematicians, and seemed a bit worried about the very idea that I would own a tractor.) But I deeply understand why people who *are* from farm families often have a very different view of their family tractor. These fine machines represent their childhood and their memories of life on the farm, of a time before the worries closed in.

△ Large areas of a tractor show are dedicated to the sale of rusty bits of metal. This is the raw material from which the beautiful restored machines on the previous pages are built. Any fifty-year-old tractor is likely to have at least some part of it missing, or maybe a bashed-in fender from when it had a heated discussion with a fence post. Online rusty parts listings have made finding the right part much easier, but there's no substitute for wandering through a few acres looking for a particularly nice specimen. Frankly, seeing these vendors makes me sad that I don't have something in need of one of these parts.

▽ Frustrated that I could not even remotely justify buying a rusty engine cowling, I bought this concrete baby deer mold instead. Then, to make it tax-deductible, I had to put it in the book.

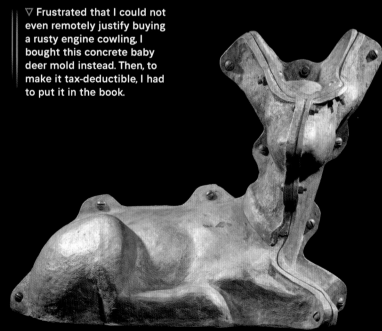

The Henry Ford Museum

THERE IS ONE NAME that represents, more than any other, the profound impact of the internal combustion engine: Henry Ford. He didn't invent the machines, but he turned them into the universal, indispensable thing they are today. He's like the Steve Jobs and Bill Gates of the car industry combined. (Steve Jobs because he made everyone want one, and Bill Gates because he was an engineer first and businessman second.)

My little statue of Henry Ford came from this machine, an interactive, automatic extrusion blow molding machine. Give it $3 (used to be 25 cents when the machines were new) and it gives you a souvenir wax model of Henry. The existence of such a machine implies the existence of something much larger: a museum for it to live in. And boy, is there a museum!

The Henry Ford museum in (where else?) Detroit, Michigan, was a personal project of Ford's, and it is as vast and earnest in its dedication to all things mechanical as was its creator. I won't bore you with details of its size, but consider this: It has the world's largest teak wood floor, spreading over $8^1/_2$ acres. That's just the indoor area; the outdoor section, called Greenfield Village, covers a further 80 acres and includes such exhibits as Thomas Edison's original workshops, moved and reconstructed from their original location in New Jersey.

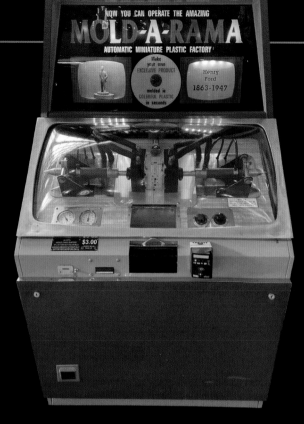

△ The automatic extrusion blow molding machine at the Henry Ford museum spits out souvenir trinkets galore.

▷ This is my molded wax figurine of Henry Ford.

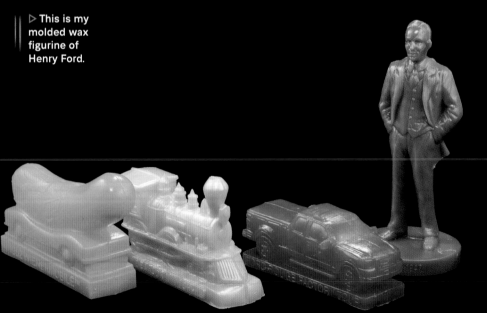

△ This is a small part of the giant teak floor at the Henry Ford Museum.

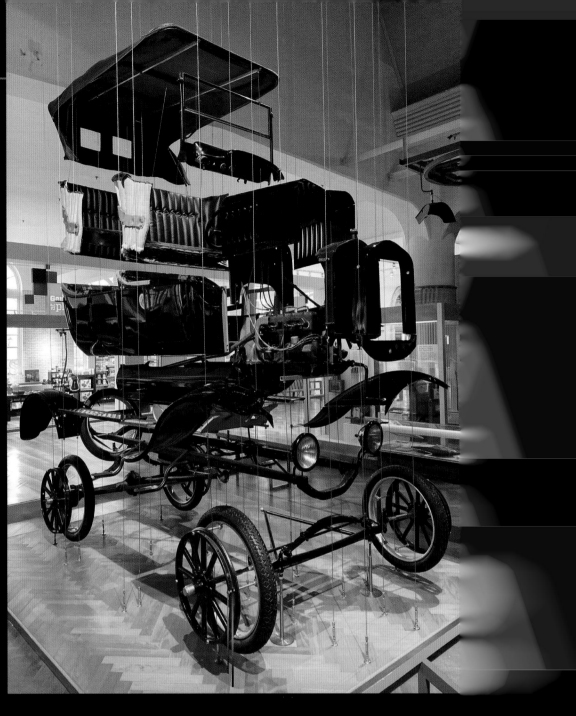

△ What do you do with 8 $\frac{1}{2}$ acres of teak floor? Well, displaying Rosa Parks's bus, George Washington's army cot, and a hot dog car are three possibilities that immediately come to mind. Elsewhere are entire steam locomotives, a dozen or more airplanes, and more cars than you can shake a piece of teak flooring at.

◁ This is why He
money to fund s
The Ford Model
produced car tha
afford. It changed
much as the inte
the same length

Cars like these represent the flashy, marketing side of Ford, but also its technical capabilities. They won races because they were fast, and they were fast because they were powerful, well-designed machines driven by people who knew what they were doing. In their time, these machines were what the best and brightest, the innovators, worked on.

▽ There are some unpleasant things you can say about Henry Ford, but there's no denying that he was the driving force behind his company right from the start, on both technical and business levels. This is the famous "kitchen sink" engine he built himself from scraps of metal and some pipe fittings, long before there was a Ford Motor Company. It's a four-stroke engine, but odd in that—as with the hit-and-miss engine—the intake valve is a simple check-valve (a valve that allows flow in only one direction, but doesn't require a cam to operate), while the exhaust valve is a cam-operated valve. Anyone can build an engine like this, and many hobbyists have. He didn't invent it, but he built it himself, and then he built a company around it that changed the world. Not many people have done that on the scale he did.

◁ It wasn't until I got home and looked at the pictures that I noticed a problem with the kitchen sink engine. The pushrod that transfers the action of the cam to the other side of the engine, where the exhaust valve is, has rotated out of position. This engine won't run until someone turns this rod about 120 degrees clockwise so the bent end points straight up. And please, someone get some oil on it! That dry metal makes my skin crawl like fingernails on a chalkboard.

Electric Motors

ELECTRIC MOTORS are superior to internal combustion engines in almost every way. They are smaller and lighter, for a given amount of power. They can be so quiet that they're essentially silent. They are remarkably maintenance free and so reliable that some can operate in harsh conditions literally for decades without failing. And many run so clean that you can operate one inside a sealed, ultrapure environment, like the inside of a hard disk, and it will not contaminate the space one bit.

So why would you ever want any other kind of motor? The problem with electric motors is that you need to supply *electricity* to run your fancy electric motor! When it comes to powering mobile vehicles like cars, trucks, and ships, the overwhelming issue is where to get enough electric power.

The amount of energy contained in today's best available rechargeable batteries is only about one-sixtieth of the energy contained in an equal weight of gasoline. When you pump gasoline into your car, you're transferring energy at a rate equivalent to several megawatts of electric power. To charge an electric car at an equivalent rate would require the capacity of a power substation large enough to serve an entire neighborhood, and that's assuming the battery could even handle such an extreme charging rate. So when it comes to powering motor vehicles for long distances, internal combustion engines still have electric motors beat by a mile—or more like a few thousand miles, in the case of a container ship.

Where electricity *is* both practical and prevalent is in stationary locations like homes, schools, and factories, as well as fixed-line transportation such as trains and inner-city buses. Here, electric motors dominate, and pretty much always have.

The average household has *way* more electric motors than internal combustion. Yes, the gasoline engine in their car is probably the most powerful motor most families own (by quite a wide margin), but it is hugely outnumbered by the vast spectrum of electric motors operating in a typical home, including refrigerator, washing machine, clothes dryer, hair dryer, shaver, PlayStation controller, electric toothbrush, hair trimmer, ear hair trimmer, mixer, blender, microwave, cell phone, garbage disposal, computer fan, ceiling fan, bathroom fan, kitchen hood fan, clocks, garage door opener, humidifier, dehumidifier, air conditioner, toy robots, actual robots, camera lenses, and on and on and on.

Which isn't even to mention the electric motors in a nonelectric car, including motors for adjusting the seats and mirrors, running the windshield wipers, pumping washing fluid, flipping up the headlights—if it's one of those cool cars with flip-up headlights—and even pumping the gasoline to feed the one single internal combustion engine in the whole contraption.

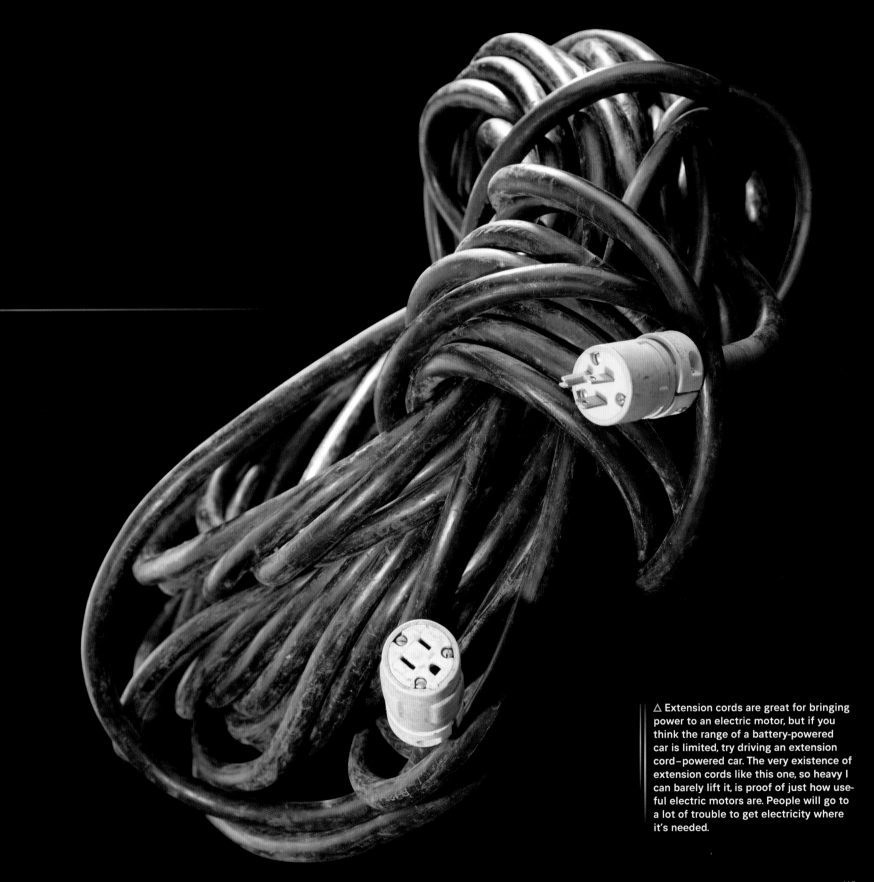

△ Extension cords are great for bringing power to an electric motor, but if you think the range of a battery-powered car is limited, try driving an extension cord–powered car. The very existence of extension cords like this one, so heavy I can barely lift it, is proof of just how useful electric motors are. People will go to a lot of trouble to get electricity where it's needed.

From Magnets to Motors

ASIDE FROM A FEW exotic curiosities, all electric motors are based on magnetic forces. Magnets pushing or pulling on each other take the place of steam pressure, or pressure from burning gasoline. So in order to understand how electric motors work, we first need to learn a bit about magnets.

There are two common types of magnets. *Permanent* magnets just sit there being magnetic all the time. *Electromagnets* are magnetic only when an electric current is flowing through them. Permanent magnets have the advantage that they maintain their magnetic field without consuming any electricity, but electromagnets have the advantage that you can turn them on and off, or flip their direction, by changing the flow of current.

To build an electric motor, you need two magnets that push and/or pull on each other. And at least one of them needs to be flipping back and forth as the motor shaft turns, to keep it going around and around. Just as with steam engine valves, if the steam always pushes in one direction, the engine will stop once the piston is all the way extended. You have to switch the direction of the force in order to pull the piston back and repeat the cycle.

There are three basic options for how permanent magnets and electromagnets can be combined to form a motor. You can have a permanent magnet spinning by a stationary electromagnet, an electromagnet spinning by a stationary permanent magnet, or one electromagnet spinning by a second stationary electromagnet.

▽ Every magnet has a north pole and a south pole. Like poles repel, unlike poles attract. In other words, the north pole of one magnet will be attracted to the south pole of another, while two north poles or two south poles will push apart from each other. What exists between and around these poles is called a magnetic field: a region of space within which mysterious forces are at play—mysterious, but easy to describe mathematically and to illustrate with what are called magnetic field lines.

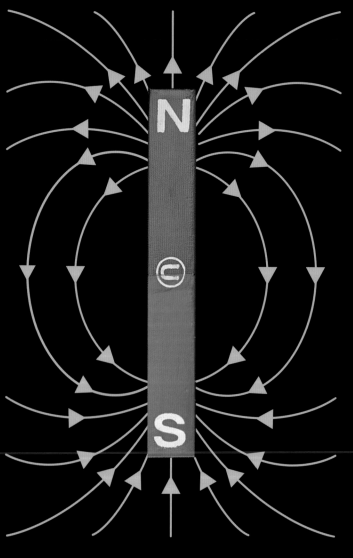

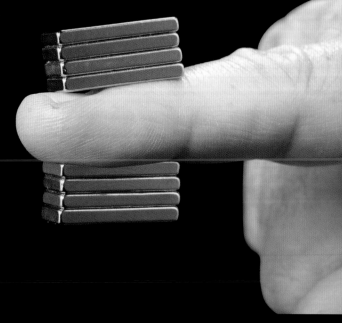

▷ Strong magnets are the single most surprising objects you can hold in your hand. If you've never held two neodymium magnets and felt them push and pull on each other over a distance of an inch or more, you really need to drop everything and go try that. Just be careful, though! Even small ones are so powerful they can crush your skin, creating blood blisters. Larger ones can crush your bones.

△ Each magnetic field line connects the north and south poles and follows a contour of equal magnetic strength. If you put a tiny compass into a magnetic field, it will point along the direction of the field lines.

▷ Iron filings line up along magnetic field lines, allowing you to see them in physical form. (There's a strong magnet under this dish.)

◁ The labeling of magnetic poles as north and south is directly related to the fact that the earth itself has a magnetic field. The north pole of any magnet always pulls gently toward the North Pole of the earth. (Ironically, this is because the earth's internal magnet has its *south* pole near the geographical *North* Pole.) Compasses are just small magnets mounted on a pivot so they can turn freely. They automatically align themselves with the magnetic field of the earth, making them incredibly useful for navigation.

What Is Electricity?

TO UNDERSTAND electric motors we need to understand magnets, including electromagnets. That means we also need to understand electricity. You've probably heard the term "electric current" quite often, or heard electricity described as "flowing" through a wire, like water flowing through a pipe. These kinds of folksy analogies to describe scientific or technological phenomena can be highly misleading, but in this case they're actually spot on, and really quite useful in understanding how electricity works. In the same way that water molecules make up the current of water that flows through a pipe, an electric current is made up of a very large number of subatomic particles called electrons. Electrons are the smallest and lightest of the three particles that make up all matter (protons, neutrons, and electrons).

Electric charges can be either positive or negative. Like charges repel and unlike charges attract. In other words, two positive charges push each other apart, as do two negative charges. But a positive and negative charge are attracted to each other.

Electrons have a negative electric charge, while protons have a positive charge. That means a collection of electrons will all try to fly apart from one another because they all have the same negative charge. But

if you mix protons into the collection they will find the electrons and pair up, attracted to each other because they have opposite charges. Electrons are often found in orbits around positively charged protons. This arrangement is known as an atom. The positive charge in the middle of the atom is known as the nucleus, and the electrons around the outside are tightly bound to it by the attraction of their opposite electric charges.

The atoms that make up metals are distinguished by the fact that they have extra "loose" electrons that can easily jump from one atom to another within the metal. These loose electrons are able to flow through solid metal just like water flows through a hollow pipe. If you pump extra electrons into one end of a piece of metal, they will push against the loose electrons inside the metal and create a "pressure" within the metal, causing the electrons to try to move away from the source of pressure.

If you make that same piece of metal long and thin, keep pumping electrons in one end, and give the electrons at the other end somewhere to go, the result is a wire with an electric current flowing through it. Which, you will note, is exactly what happens when you pump water into one end of a pipe.

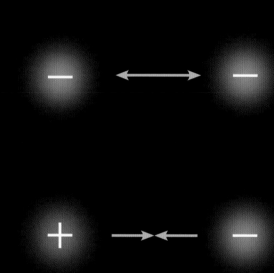

▽ In the case of water, you have water molecules pushing on each other to transmit pressure from one end of the pipe to the other. Water flows in a direction from higher water pressure to lower. If you increase the pressure, more water will flow through the same pipe. If you keep the pressure the same but increase the diameter of the pipe, again more water will flow.

▽ Voltage is the equivalent of pressure for electrons. In a wire you have electrons pushing on each other to transmit voltage from one end to the other. Electrons flow in the direction of higher voltage to lower. If you increase the voltage, more electric current will flow through the same wire. If you keep the voltage the same but increase the diameter of the wire—or make it out of a lower-resistance metal—more electric current will flow.

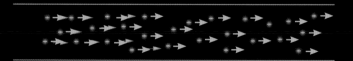

The Connection Between Electricity and Magnetism

THERE IS A DEEP, bi-directional connection between electric currents and magnetic fields. Anytime you have a wire with an electric current flowing through it, there will be a magnetic field surrounding that wire. If the wire is straight, the field will be wrapped around the wire, as if you had a ring magnet with its north pole wrapped all the way around to meet up with its south pole.

If you wind the wire into a coil, all the circular fields formed around each length of wire join together to form a magnetic field with a north pole at one end of the coil and a south pole at the other end. If you compare these field lines to those of a permanent magnet, you'll see why this device is called an electromagnet. It behaves in every way like a permanent magnet, except that the magnetic properties vanish when you switch the current off. And if you reverse the direction of the current, the north and south poles will be on opposite sides. This is the key characteristic that lets you make motors out of electromagnets.

Currents create magnetic fields, but magnetic fields can also create currents. If you move a magnet past a wire full of electrons, it will create a force on each electron, pushing it through the wire. You can think of it like a magic water pump that pushes on the water in a pipe without actually touching it. This is what makes electric generators and transformers possible.

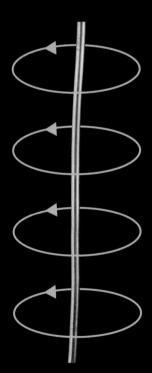

△ A straight wire creates a circular magnetic field.

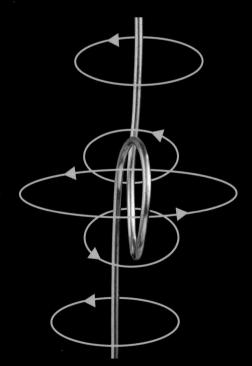

△ A circular wire creates something more like a straight magnetic field.

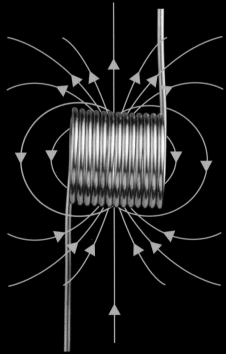

△ A whole coil of wire creates a magnetic field very much like that of a permanent magnet.

▽ This simple electromagnet—a coil of wire wrapped around a screw—will pick up small pieces of iron, but only when an electric current is flowing through it. (In this picture the current is coming from a battery.) The screw (or any iron core) concentrates the magnetic force on its ends. As soon as you cut the current, the piece drops. Just as a permanent magnet will pick up unmagnetized iron with either end, this electromagnet will pick up iron objects regardless of which way the current is flowing.

▷ This drill is meant for making holes in steel I-beams that are too large to put under a drill press (or are already installed in a building) and too thick to be drilled with a handheld drill. It uses a powerful electromagnet to clamp itself magnetically to the I-beam, allowing the operator to apply a great deal of pressure to the drill bit using the revolving handle.

These buttons turn the electromagnet on or off, or engage its "demagnetize" feature.

These buttons turn the drill on (in forward or reverse) and off.

From the bottom we can see the iron core of the electromagnet. The copper coils are hidden by a potting compound poured around them to protect them from moisture or physical damage.

This spring is holding the iron rod out away from the coil.

With voltage applied to the coil, the rod is pulled in, compressing the spring.

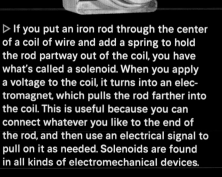

▷ If you put an iron rod through the center of a coil of wire and add a spring to hold the rod partway out of the coil, you have what's called a solenoid. When you apply a voltage to the coil, it turns into an electromagnet, which pulls the rod farther into the coil. This is useful because you can connect whatever you like to the end of the rod, and then use an electrical signal to pull on it as needed. Solenoids are found in all kinds of electromechanical devices.

Sorry for the dirty fingers.

▷ If you use a solenoid to move the contacts of an electrical switch, you get what's called a relay. This lets you use one electrical signal to control another. So, for example, you could arrange it so that when voltage is applied to the coil, the contacts close a circuit that is feeding a much higher voltage, or much more current, to another device. Or, you could wire it so that when the voltage is turned *on* in the coil, this *disconnects* the contact, shutting *off* voltage somewhere else. In computer terms, this would be called an inverter, one of the fundamental operations of digital logic.

The movable contact is touching the top contact.

◁ With no voltage supplied to the coil, the movable contact is held in the up position by the spring on the right.

Now the movable contact is touching the bottom contact.

◁ With the coil energized, the movable contact is pulled down against the core of the coil.

▷ In more complicated devices one solenoid can control more than one contact, or two solenoids can work together to make a sort of "memory." This compound relay can latch in one of two states: When solenoid on the left is momentarily activated, the contacts are locked in one state and stay there. If the solenoid on the right is activated, the contacts latch in the other state. In other words, this device will "remember" which coil was last activated, even when there's no more current flowing to it. It is, in effect, storing one binary digit—one "bit"—of digital information. If you hook up a *lot* of relays together, you can build an entire computer that carries out calculations by opening and closing relay contacts. In fact, between about 1930 and 1950 relays like this were the main way computers were built. Then vacuum tubes took over, followed by transistors. Both are like relays, in that they use one electrical signal to control another, but they are faster, more reliable, and less power-hungry.

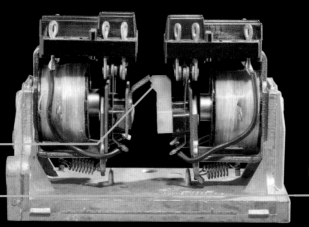

This mechanism creates the latching effect.

The contacts can be latched either to the left or right, connecting two different sets of terminals.

This free-swinging magnet is the secret to how the relay latches. If the coil is energized in one direction, the magnet is pulled straight toward the coil. When the coil is turned off, the magnet blocks the solenoid rod from raising up again, trapping the relay in its "on" state. If the coil is energized in the other direction, the contact plate is again pulled down, but the magnet—because it's a magnet and not just a piece of iron—is repelled by the field from the coil. The magnet swings out of the way, allowing the iron rod to move up and the contact to open when the current is shut off.

Negative Charge

My description of electrons flowing from higher to lower pressure (higher to lower voltage) is deceptive in an unfortunate way. In a logical world, the physical particles called electrons would flow from the side of high, positive voltage to the side of low or negative voltage. If each electron carried a positive charge, this is exactly what would happen. But—and this is entirely Benjamin Franklin's fault—electrons actually have a negative charge, and they flow from the negative side to the positive side. When Mr. Franklin figured out that electricity was a flow of particles, he had no way to tell which direction the particles were flowing, and thus what charge they had. He simply had to guess, and he guessed wrong. By the time we were finally able to figure out how to determine which way the particles were flowing, it was too late to change the convention, and we've lived with Franklin's mistake ever since. Fortunately, it doesn't matter in nearly all situations. By convention, "electric current" is said to flow from the positive side to the negative side, even though it's actually electrons flowing in the opposite direction. The math all works, so we just don't worry about it.

△ Confusingly, in the automotive world people often use the word "solenoid" to refer to a relay. For example, if your mechanic tells you that you need a new starter solenoid, they actually mean you need a new relay. This is a cool little latching relay, called a battery disconnect solenoid, that I had to replace on a camper my daughter was fixing up. If you connect the voltage to the coil in one direction, the contact will close and stay closed even after you turn off the voltage. But if you connect the voltage the other way (so the magnetic field is reversed), then the contact will open and stay open after the voltage is switched off again. This lets you switch the relay on or off with just a pulse of electricity on the coil, after which it stays the way you put it without using any power in the coil.

This copper ring makes a connection between the terminals on the left and right sides, when it's pressed down onto them.

This solenoid coil pulls the contact plate down when voltage is applied to it in either direction. Because the core of the solenoid is iron, it's attracted to the magnetic field regardless of which way the field is pointing.

This iron rod extends down into the solenoid coil and is pulled down when the coil is energized.

These contacts and terminals are large because they are designed to carry up to 100 amps of current—which is a lot.

From Electromagnet to Motor

OW THAT WE'VE LEARNED the
sics of the science of electricity
d magnetism, we can turn to the
gineering of making a motor that
tually works. We begin with the
me coil of wire wrapped around a
il from the last section.

If we point this coil at a piece of
dinary iron, it will attract the iron
gardless of which way the current
flowing. But if we point it at a per-
anent magnet, it will either pull
push on the magnet, depending
the polarity: North poles attract
uth poles but repel other north
les, and vice versa. If we put the
agnet on a swivel and wire the coil
a switch that can flip the direction
the current, we can force the coil
alternately pull one or the other
d of the magnet toward it.

Already you can see how this
ight turn into a motor. Switching
e direction of the current back
d forth will cause the magnet to
p back and forth. If the timing is
st right, the magnet should begin
spin. Then we need to automate
e process. Think back to the way
steam engine works: we'll need
e equivalent of the steam valve, or
me way of automatically reversing
e direction of the current at just
e right moment to keep the mag-
t spinning.

With the electric-
ity flowing in this
direction, the north
pole of the magnet
is attracted to
the coil.

With the flow of
electricity reversed
the south pole of
the magnet is at-
tracted to the coil.

The Simplest Electric Motor

THE EASIEST WAY to achieve the automatic switching of the electric current is to switch the status of the magnet and the coil. Instead of a fixed coil and a rotating magnet, we have a fixed magnet and a rotating coil. A pair of sliding electrical contacts, called a commutator, are arranged in such a way that the current automatically changes direction at just the right moment.

You could make a motor with just one coil and one magnet, but it's much more common to have several pairs of magnets and coils, which makes the motor stronger and smoother-running. In this example there are two coils and two magnets on opposite sides, but you can have as many as you like arranged around a circle. Some motors have a hundred or more!

The arrangement shown here is very common in electric motor models, displays, and educational explanations of how electric motors work. But you won't find any real motors that work in exactly this way, for a couple of reasons. Just as in a single-cylinder steam engine, there's a dead spot when the coils are facing directly at the two magnets. If the motor is stopped in this position and you apply power to it, there won't be any net force in one direction or another, and the motor won't start. Worse, anytime the motor is rotating through this position, the stationary contacts (called brushes) will briefly touch both sides of the split ring contacts, resulting in a short circuit through the contacts from one brush to the other. There is a solution that is both obvious and surprisingly not obvious: use three coils instead of two, as we will see on the next page.

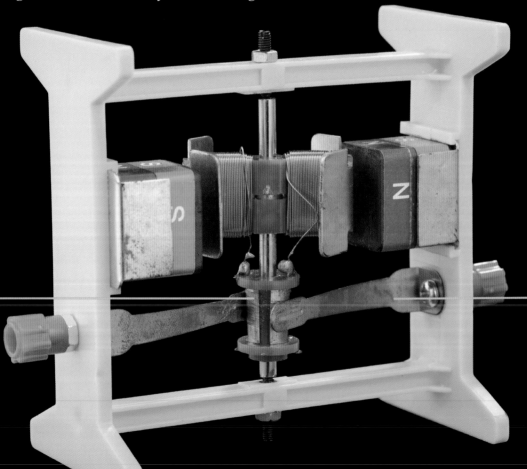

These contacts route current from the copper strips up through the wire coils. When the rotor has made half a turn, the position of the two rotating contacts is switched, and the current starts flowing in the opposite direction.

Current flows through these copper strips, called brushes, which are bent to press lightly against the rotating part. Together with the contacts they are touching they form a commutator.

The Simplest *Good* Electric Motor

AS EINSTEIN FAMOUSLY SAID, you should make things as simple as possible, but not simpler. The motor on the last page with two coils is just a bit *too* simple. But if you add one more coil and contact, bringing the total to three each, something amazing happens.

Three coils and three contacts allow us to avoid both a dead spot, and any position in which there is a short circuit through the contacts. When power is applied to the motor, it is guaranteed to start, and it will always turn in the same direction.

The three contacts ensure that each coil is always energized in the right direction to keep pushing the motor around. The current is cleverly split, some of it flowing through a single coil while the rest of it flows through the other two coils in series. Imagine a triangle: Each face of the triangle is one of the three coils, and each corner is one of the three contacts. If you connect a voltage across two of the corners of the triangle and leave the third corner unconnected, current can flow both through the single side between those two corners, and through the other two sides in series.

I don't know who discovered this, but I'm sure the moment they realized that this arrangement would actually *work* must have been one of the most satisfying moments of their whole life. It's not at all obvi-

ous that an arrangement with these properties has to exist, but it does, and it's what you'll find in billions of motors in use today. It's not the most sophisticated or the most efficient, but it's by far the cheapest to make, and on the whole it works really, really well.

▽ The coils in this motor are wired in a triangle, with current applied to two of the corners at a time. The flow of current is split: Two-thirds of it flows through the more direct route along the edge connecting the two corners where current is coming in. The other one-third flows through the longer route along the other two edges in series.

△ This model shows the arrangement of coils, magnets, and contacts used in billions of basic three-pole DC electric motors, from tiny toy motors to powerful starter motors in cars.

The commutator in this model has two brushes but *three* rotating contacts, one for each coil.

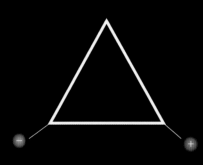

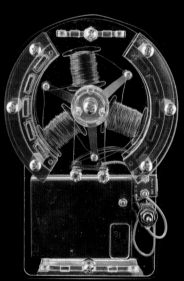

△ My version of this style of motor is designed to let you see how the coils are wired in a triangle configuration. Two brushes touch the three rotating contacts, each of which is connected between two of the coils.

THE PATTERN OF current flow in a three-pole motor is both subtle and clever. How does everything fit together so that the polarity of all three coils is always exactly right to help push in the right direction? I didn't really understand it until I designed and built this model and played with it for several minutes. *Then* it all made sense. Let's have a look.

This model has the same commutator and brushes as a real motor, but instead of coils, it has pairs of blue and red LEDs that light up to show where the north (red) and south (blue) poles are at any given moment. One will always be red and the other blue, to indicate the direction of the magnetic field being created by that coil. It's the LED on the outside that matters, because that's the end of the coil that is close to the permanent magnets around the outside.

The brightness of each LED also shows whether current is flowing directly across that "coil" from one commutator contact to the next, or indirectly through two of them in series.

If you spin the model quickly, all the lights blur together to form a pattern that reveals the secret.

At the exact moment when a coil reaches the point where it's pointing directly left or directly right (at west and east positions if you think of it as a compass), the coil reverses polarity,

▷ Look at the schematic diagram and follow the flow of current with your eyes. It enters through the red brush, which is touching the commutator contact that points down and to the left. The green brush on the right is touching the contact that points to the right. From these two contacts, current can flow through two paths: directly through coil 1, or through coils 2 and 3 in series. The contact pointing up is not touching anything, so it's just along for the ride.

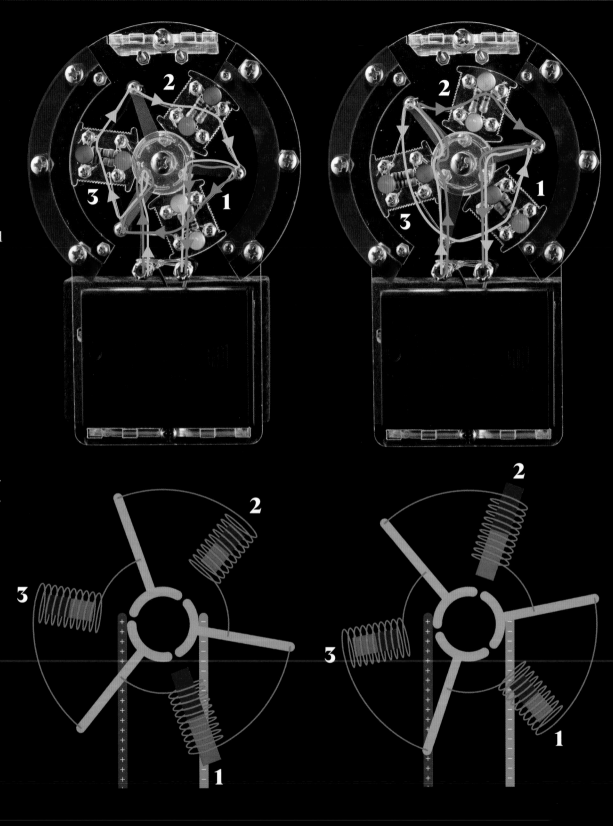

switching from blue to red and from having the opposite polarity to the magnet it's near (pulling in) to having the same polarity (pushing away). The LED is dim both before and right after the coil switches polarity. That's because it's being powered by current flowing through two coils in series. The contact on the opposite side has stayed connected to the same side of the battery, while the brush on the same side has switched from one contact to the next.

Each coil stays the same polarity for half of a revolution, always red on top and blue on the bottom. But for the center third of this half-turn, the LED is bright. That's the time during which it is connected directly between the two sides of the battery and is getting full-strength current. As an added bonus, this is also the time during which a strong magnetic field from this coil is most useful for powering the motor.

I really can't overstate just how clever this is. Gold star for whoever invented this thing.

◁ In this position, rotated just a little bit counterclockwise, the brushes have switched to the next set of contacts, and again all three coils are energized, but now it's coil 2 that is running at full strength, while coils 1 and 3 are running in series with each other. Coil 3 has switched direction, because current is now flowing through it in the opposite direction. As if by magic, all the poles are again pointing in the right direction to assist in keeping the rotor going counterclockwise. As the motor is turning, each time it is *just about* to reach the point where the poles of the coils would start pushing in the wrong direction, the contacts switch, the current flow shifts to a new combination of coils, and the poles switch direction to keep pulling counterclockwise.

▷ The three-pole design is very commonly used in the small DC motors found in toys or anything else that requires a cheap, simple motor that runs on low-voltage battery power. This is one of the motors that I use in many of my model designs to animate their movements. It's the simplest and least expensive practical design for a DC motor.

This type of motor runs most efficiently at high rpm. If you want something slower, you need a gear box to cut down the speed. This one gives a final output of about 100 rpm (about 1.6 revolutions per second).

I accidentally bent this brush taking it apart.

The tiny coils are wound with hair-thin, lacquer-insulated magnet wire, soldered to the three commutator contacts.

More serious motors have graphite brushes (see below), but this tiny toy motor appears to have brushes made of tinned copper. That means its lifetime is limited to when friction wears down, and eventually through, the thin copper sheet.

▷ Here's the toy motor pretending to be the starter motor for one of my internal combustion engine models. Real car starter motors are basically the same type of motor but made with a *lot* more copper.

▷ This is an actual starter motor. Its job is to start up a much larger gasoline or diesel engine. What you see here is actually an assembly that includes not only the motor but also a solenoid, a planetary gear reduction stage, and a small gear that engages with gear teeth around the outside of the engine's flywheel.

This gear is normally withdrawn into the housing of the starter motor, but it pops out, with the help of a powerful solenoid, when it's time to start the motor. Once the engine has started, hopefully just a few seconds later, the gear pops back in, allowing the flywheel to turn freely.

△ The housing of the starter motor assembly is a beautiful example of die casting. It's a single, very complicated piece of metal made by pouring molten aluminum or zinc alloy into a multipart steel mold, which pulls apart after the metal has cooled. Millions of these are made every year at very low cost.

△ Starter motors aren't very big, only about 6 inches (15 cm) long, but they draw a tremendous amount of current (200 amps or more) for a few seconds at a time, and deliver enough torque to turn over a large gasoline engine. Drawing 200 amps at 12 volts, this motor consumes 2,400 watts of power, more than the average used by an entire house. But only for a few seconds.

Everything about this motor is designed to handle huge electric currents. These carbon brushes, which make the sliding electric connection to the rotor, are thick and wide, and there are two separate sets of them. Only one pair is required but having two allows twice as much current to flow.

The braided copper wires carrying current to the brushes are about $1/4$ inch (5 mm) thick. That's a lot for a wire.

Strong coil springs press the brushes hard up against the contacts in the commutator. This creates a lot of friction, but it's nothing compared to the amount of torque this motor is expected to generate. The pressure helps make a solid connection able to handle a high current flow.

▷ The first big, powerful electric motors, which date back over a hundred years, were DC brush motors of basically the same design we just described. For decades there were DC (direct current) electric distribution systems in cities specifically to bring DC electricity for motors like this. As we will see later in this chapter, the world switched, for very good reasons, to AC (alternating current) distribution, and many large industrial motors long ago switched to different designs that avoid brushes.

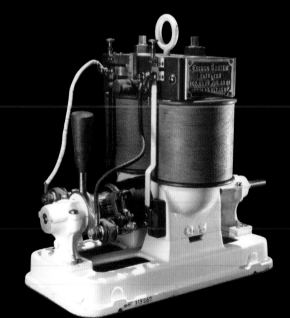

It's difficult to make wire coils completely consistent. There's always some slight variation among coils, which makes the rotor not perfectly balanced and causes vibrations when it's spinning at high speed. To fix that, each motor is individually tested and balanced by cutting away just the right amount of material in strategically located slots, much like car tires are balanced with little weights attached to the wheel rims.

The coils are made of thick, heavy copper wire able to carry hundreds of amps, at least for a short time. They're coated in a layer of glue that locks them in place. This glue is more important than it might seem. All the mechanical force that turns the rotor is created by magnetic forces *inside these wires*. The wires push on the rotor, not the other way around. If they came loose, they would bend and rub until they wore a hole in their insulation, shorting out the motor.

With the brushes removed, we can see the thick copper contacts they slide against. Copper is a very good conductor of electricity. It's also expensive, so you can be sure that this generous amount is necessary to handle the current.

Surrounding the rotor are four strong rare-earth magnets, hidden inside a stainless steel casing.

▷ The brushes in serious motors are always made of graphite, and this brush is especially serious given how *big* it is; most are only about half an inch (1 cm) long. Graphite is really the only material that meets the requirements. It conducts electricity, slides smoothly over metal without lubrication, and can withstand electric sparks without melting. Graphite powder is actually used as a lubricant, and graphite is so temperature-resistant that it is used to make molds for casting high-melting-point metals. But despite being a nearly ideal material, graphite brushes do eventually wear out and require replacement. And while they are in use, they generate fine graphite dust, and a lot of audible and electrical noise as they spark against the commutator contacts. You could never, for example, use a brushed motor inside a hard disk drive, where everything must be kept absolutely clean.

Brushless DC Motors

THE SIMPLE DC BRUSH motors on the previous page are very common, and this was the first kind of electric motor to be widely used for serious applications. But these motors have some problems, mainly centered on the brushes.

Electricity likes to flow through solid metal, not jump across gaps; a sliding electrical connection is never a good thing. To give the electricity a solid wire connection to an external source of power, the coils need to be stationary. That means the permanent magnets need to be spinning on the motor shaft. This is no problem, as it makes very little difference to the operation of the motor whether it's the coils or the magnets that are spinning—but if the magnets are spinning, we still need some way to make the current in the coils flip back and forth at just the right moment. In other words, we need some way of sensing the position of the motor shaft, and some way to use that information to control the current in the coils.

There are several ways of detecting the position of a rotating shaft. These days, the most common way is through a magnetic field sensor called a Hall effect sensor. It's mounted near the rotor and sends a signal each time one of the magnets in the rotor passes by the sensor. The signal tells an electronic controller (built with solid-state semiconductor components) to reverse the polarity of the current going through the coil, which keeps the motor spinning. The combination of sensor and controller circuit takes over the function of the brushes and contacts in the commutator of a brush motor.

This arrangement is called a brushless DC motor. It's made possible by the existence of small, cheap, reliable semiconductor chips: basically the same technology used to make computer chips and cell phones, but applied to controlling the flow of larger amounts of electricity.

△ This model brushless DC motor shows the main components very clearly: a fixed coil, rotating magnets, and the Hall effect sensor.

A stationary coil pushes and pulls on the rotating magnets.

A simple electronic circuit acts as a commutator to reverse the direction of the current in the coil.

The Hall effect sensor detects the presence of a magnetic field every time a magnet comes around.

▽ Computer cooling fans are a common application of brushless DC motors. They run on 12–volt DC and can easily run without stopping for years on end, maybe even decades, without maintenance. This kind of reliability is possible because there are no moving mechanical parts whatsoever, except of course the spinning fan and rotor. There aren't even any ball bearings, because the motor's magnets themselves hold the rotor and fan in place. There is only very light pressure on the sleeve bearings at the ends of the shaft. To be more technical, no thrust bearing is necessary, because the force generated by the blowing air is countered by magnetic force, not a physical bearing. High-end models actually use "air bearings," which are machined so precisely that they spin without friction on a cushion of air. And of course there are no electrical contact points, just solid-state electronics that last nearly forever (as long as they aren't subject to voltage spikes).

This inexpensive fan has a relatively low-precision greased bearing that will probably only last a few years.

Hall effect sensor

Ring of permanent magnets embedded in a rubber strip

◁ The CompactFlash card format was popular in cameras and camcorders for several years before it was superseded by even smaller formats.

▽ Inside this tiny device, you can see something that I did not believe was possible when I first heard it was being made: a 1-inch (2.5 cm) diameter hard disk.

Stationary coils

The tiny, ultra-flat aluminum disk, coated with iron oxide, spins at high speed.

Rotating permanent magnets

An arm holds tiny electromagnets (one on each side) that slide back and forth over the surface, reading and writing data to tiny magnetic zones—16 billion of them on each side.

△ This is an example of a technology reaching a ridiculous level of refinement *just* before it becomes completely obsolete and vanishes from the face of the earth. It's a mechanical hard disk built into the form factor of a Compact-Flash card, a format intended to house solid-state flash memory for use in cameras. For a brief moment in time, it was possible to get more capacity with a miniaturized hard disk than by using actual flash memory. Today you could fit a thousand times as much memory in the same space using higher-density flash memory chips.

△ Disk drives typically run off the 5-volt or 12-volt power supply in a computer, but putting a brushed DC motor inside the sealed, dust-free case would be nuts: graphite dust from the brushes would destroy the platters in seconds. You could put the motor outside the housing, but then you would need the rotating shaft to go through the wall of the sealed box that keeps the platters clean and protected. The solution is a brushless DC motor built into the doughnut hole in the middle of the platters, or at least flat up against them.

△ The inside of a camera lens is another place you wouldn't want a brush motor. These tiny brushless motors operate the focus and aperture mechanisms for a variety of (broken) lenses.

▽ A brushless DC motor must have some way of measuring the position of the rotor, so it can switch power to the coils at just the right moment. Hall effect sensors are common, but the most accurate way of measuring the position is with an optical shaft encoder. This device has a tiny LED that shines light through a slotted wheel toward an optical sensor. As the wheel turns, the slots break the beam, and by counting the pulses from the sensor you can determine the position very accurately—more accurately than is typically necessary just for powering the motor, but useful for more sophisticated controllers that adjust the power in order to very precisely control the rotation of the motor. The slots count only relative motion; to get a starting point from which to count, there is usually a separate limit switch that tells the machine when it has moved all the way to one end of its range.

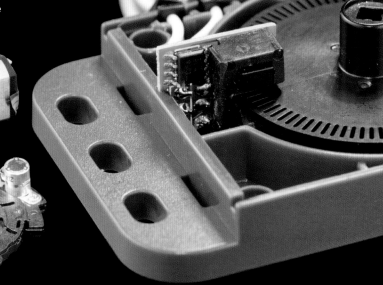

△ Modern cordless power tools often have brushless DC motors. Some manufacturers of these tools are particularly proud of this fact and advertise it on the side of the motor housing. Some even go so far as to print the word "Brushless" *bigger* than the company's own brand name! Fair enough, these motors are better than the older brushed design, but they could just as well have picked "Neodymium" as the slogan to run with, since the use of ultrapowerful neodymium rare-earth magnets is at least as important a factor in the superior performance of these newer tools. And let's be honest, which sounds cooler as an advertising hook, "Brushless" or "Neodymium"?

▷ This is an amusing toy motor that combines the idea of an optical shaft encoder with solar cells. Instead of an LED and a light sensor, it uses the sun and power-generating solar cells wired directly to its coils. Instead of a mechanical commutator or electronic controller, it uses *shade* to switch the current on and off. As the motor is running (quite fast in bright sunlight), the black wedges alternately cover and uncover different pairs of solar cells, each of which is wired to a separate pair of coils. The exposed coils are always just a bit ahead of the next magnet around the circle, so the motor is always pulling itself farther around. There is no circuitry in this design: no commutator, no motor controller, no electrical contacts, no nothing. Just solar cells permanently wired to coils of wire. It will run literally as long as the sun shines!

▷ This motor is an even simpler version of a shade-commutated motor. Instead of vanes that cast shade, solar cells are attached directly to the rotor, so they turn themselves away from the sun as the motor is going around. As in the toy motor on the previous page, the solar cells are wired directly to the coil of wire in the rotor, with no commutator, controller, or electrical contacts.

Solar cells collect sunlight and convert it to electric current—but only when they are facing the sun.

Current flows through whichever of these coils has its corresponding solar cell facing the sun.

Permanent magnets create a magnetic field that the coils can push against.

△ This horizontal design of a solar cell motor is commonly known as a Mendocino motor, because it was invented in Mendocino, California. It's exactly the same as the vertical version above, except inferior for my purposes because you can't see the coils and how they are wired to the solar cells. They are hidden inside the body of the motor. These solar motors all use magnetically levitated bearings, because normal ball bearings would have too much friction. They are very weak: They barely generate enough power to keep themselves moving, let alone drive a useful load. This isn't to imply that solar cells can't generate enough electricity to do useful work; you just need a whole lot more of them than fit in the space of a small motor.

AC Motors

WE STARTED THIS chapter with the analogy that electricity flowing in a wire is a lot like water flowing in a pipe. Up to now we've been discussing "direct current," or DC, which is the current, always flowing in the same direction, that you get from a battery or power supply. The water analogy works well when we're talking about DC, because water in a pipe likes to keep flowing in one direction. But there is one thing you can do with electricity that you can't do with water: change the direction of its flow back and forth many times per second.

This is known as "alternating current," or AC, and it's the type of current you get from any standard wall outlet. If you combine the two types of current you get AC/DC, which is an '80s rock band that has nothing to do with electricity.

Common household power in the United States is AC that switches back and forth 60 times per second. In other countries it's always either 50 or 60 times per second. Airplanes have standardized on 400 cycles per second for their internal electrical systems.

You may have noticed that much of the discussion of DC motors so far in this chapter has focused on various ways of switching DC current back and forth in order to make a coil alternately push and pull on a magnet. In other words, we were talking about how to turn DC into some form of AC. What if we just started with AC? This is in fact a great idea, and as a result the majority of larger electric motors in the world are AC motors.

To properly understand AC motors, we need to understand how AC can be turned into a *rotating* magnetic field.

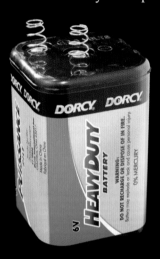

△ Power from a battery is DC. The voltage stays the same all the time.

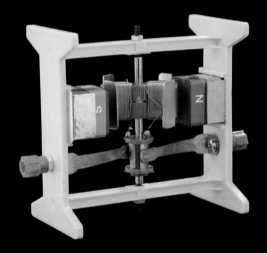

△ The power going to the coils in the DC motors we looked at in the last section switches suddenly from one direction to another when the contacts rotate and connect to the DC power source in the opposite direction. This creates a "square wave" of alternating positive and negative voltage going to the coils.

△ The voltage from a wall doesn't just switch suddenly from positive to negative; instead, it goes up and down in a smooth curve called a sine wave. Later we will learn about the close relationship between sine waves and circles, which turns out to be central to the function of AC motors.

Rotating Magnetic Fields

KEY TO THE OPERATION of AC motors is the rotating magnetic field. Imagine two magnets facing each other. One magnet, the input magnet, is attached to a crank, and the other, an output magnet, is connected to a gear. The two magnets are attracted to each other, north pole to south pole, so when the crank rotates the input magnet, the output magnet will follow, turning the gear connected to its shaft. The magnets are not touching each other, so they are connected only by their magnetic fields. You could say

that the output magnet is being rotated by the *rotating magnetic field* created by the input magnet.

The rotation of the output magnet is *synchronous* with the rotation of the input magnet. It will always turn at exactly the same speed. If there's some drag on the gear, the magnets will start to separate a bit, which increases the force exerted by the input magnet (right up until the point where the force is too great and it slips by a whole turn).

If, on the other hand, you push on the gear to try to make it go faster than the crank, the input magnet will actually pull it backward, trying to slow it down to keep it going at the same synchronous speed. Within the limits of the strength of the magnets, the two are trapped together, and any attempt to get them to turn at a different rate results in a force that tends to pull them back into alignment.

Arrangements exactly like this are in fact found in a number of useful devices.

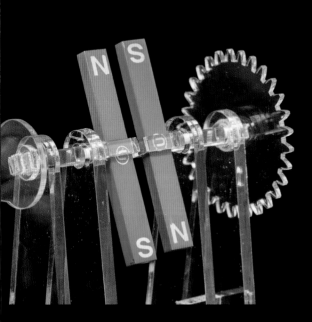

△ In this demonstration of a rotating magnetic field, these two magnets are magnetically locked to each other.

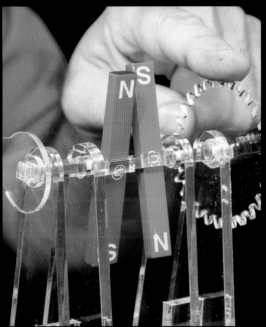

△ Friction on the output magnet causes it to lag behind the input magnet, but the two keep turning at exactly the same speed.

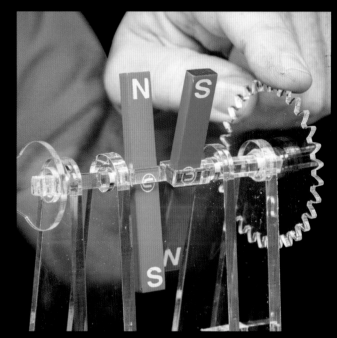

△ Try to make the output magnet go faster, and you get a force pulling it back. Again, it keeps turning at the same speed as the input magnet.

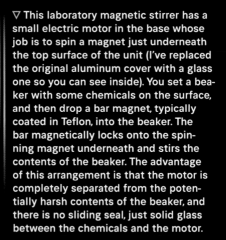

▽ This laboratory magnetic stirrer has a small electric motor in the base whose job is to spin a magnet just underneath the top surface of the unit (I've replaced the original aluminum cover with a glass one so you can see inside). You set a beaker with some chemicals on the surface, and then drop a bar magnet, typically coated in Teflon, into the beaker. The bar magnetically locks onto the spinning magnet underneath and stirs the contents of the beaker. The advantage of this arrangement is that the motor is completely separated from the potentially harsh contents of the beaker, and there is no sliding seal, just solid glass between the chemicals and the motor.

▷ Inside a magnetic stirrer, there is literally a motor turning a magnet.

Magnet

Motor

ehm
M1G055-AI01-01
24V-(16-28V-) DC
4W 1000U/min
Electronically protected
Made in Germany
24/03

A Teflon-coated bar magnet goes in the beaker of harsh chemicals.

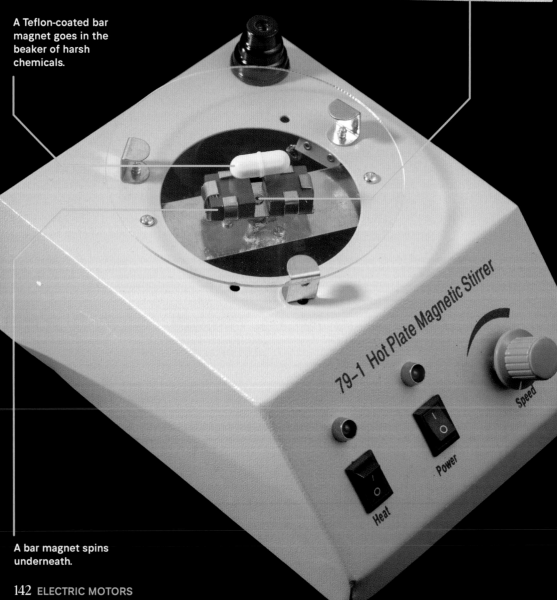

79-1 Hot Plate Magnetic Stirrer

Speed

Power

Heat

△ Magnetically coupled impellers are not just for beakers of acid and toxic waste pumps. This battery-powered snowman lantern—with glitter snow and LED lighting—has exactly the same mechanism inside.

A bar magnet spins underneath.

A small DC brush motor in the base spins a pair of permanent magnets.

▽ The same idea of magnetic coupling is used in pumps like this one (which I have cut open to show the parts inside). The impeller (like a fan blade for water) is not physically connected to the motor. It is completely isolated inside the pump housing, without any shaft passing through to drive it. Instead, it is magnetically locked to a rotor that is spun by the motor. The great advantage of this design is that there is no need for a watertight sliding joint where the drive shaft from the motor would otherwise have to enter the pump housing to drive the impeller. Sliding watertight joints, like sliding electrical connections, do not spark joy.

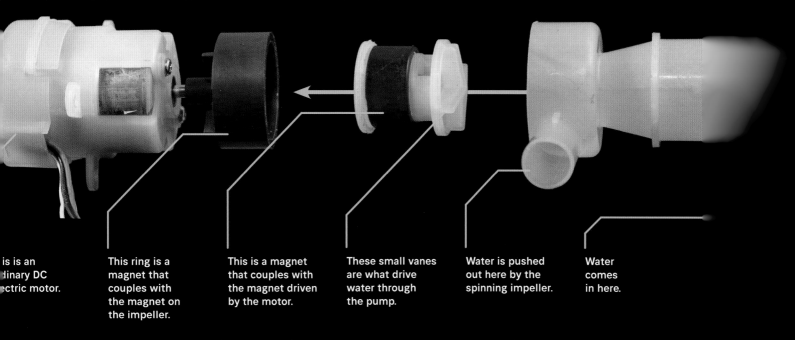

is is an
dinary DC
ectric motor.

This ring is a
magnet that
couples with
the magnet on
the impeller.

This is a magnet
that couples with
the magnet driven
by the motor.

These small vanes
are what drive
water through
the pump.

Water is pushed
out here by the
spinning impeller.

Water
comes
in here.

▽ At first this water pump looks very much like the one before, with a magnetically coupled impeller. But look closer: There's no motor and spinning magnet to drive it. Instead, there are just some coils of wire. This is, in effect, a brushless DC motor that has a solid plastic cup fitted between the permanent magnet rotor and the stationary coils. But it also hints at the ide
can, if properly arranged and powered, create a rotating mag
make a nearby permanent magnet spin, which is the princip
type of motor we're going to learn about.

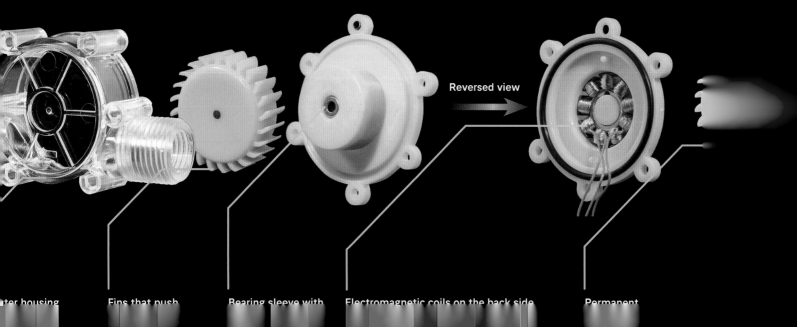

Reversed view

ter housing

Fins that push

Bearing sleeve with

Electromagnetic coils on the back side

Permanent

Rotating Magnetic Fields Without Rotating Magnets

AS WE JUST SAW, using one permanent magnet to turn another can be useful, but it's not useful *as a motor*. What if we could create a rotating magnetic field using stationary electromagnets (coils of wire) instead of a rotating magnet? Then we could use that rotating field to turn a permanent magnet, and *that* would be a motor.

We need to consider how magnetic fields from multiple sources interact with each other. If you hold two magnets together pointing in the same direction, you'll get a magnetic field that's twice as strong. If you hold them at an angle to each other, you'll get a field that points in a direction in between, like an average of the two directions.

I made a device to show how this works. It's got a permanent magnet on a swivel in the middle, so we can see the direction of the field at the center. And it's got two slots around the outside where I can put one or more magnets. By sliding these magnets in and out, I can vary how much each one contributes to the field in the middle.

With just one magnet in place, the field points in the direction of that magnet. If I start to slide a second magnet slowly into place, the field starts rotating in the direction of the new magnet, and once it' all the way in, the field is halfway betwee the two. Remove the first magnet and the field will now point in line with the secor magnet, having made a smooth 90-degre turn. We could repeat this process with a magnet in the third position, then the fourth position, and then back to the first position, giving us one complete revolution of the field in the center.

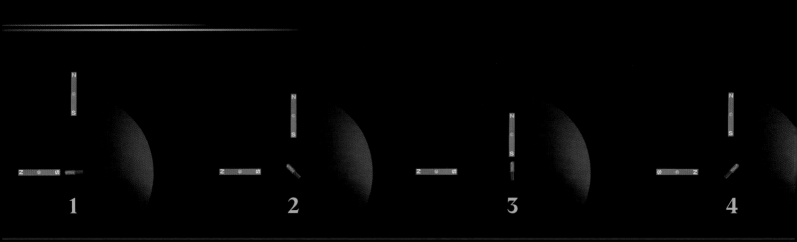

STEP 1 With the horizontal magnet in place with its south pole facing in, the field points in line with that magnet and the north pole of the rotor is pulled to the left. (The other magnet is far enough away that it has no effect.)

STEP 2 With both magnets partially in, both with their south poles facing in, the field points midway between them, rotated 45 degrees clockwise from the previous step.

STEP 3 Now we've moved the horizontal magnet all the way out, and the vertical magnet all the way in. The field has again rotated 45 degrees clockwise, and now faces straight up.

STEP 4 Here both magnets are partially in, but we've flipped the first one so its north pole faces in. This pushes the north pole of the rotor away, achieving another 45-degree clockwise rotation.

STEP 1 At this step the horizontal magnet has its south pole all the way in, so its data point (green) is at the top of the graph (positive position). The vertical magnet is out, so its data point (red) is in the middle (zero point).

STEP 5 Here the horizontal magnet is all the way in, but with its north pole facing in, so its data point (green) is at the bottom of the graph (negative position).

Here's what it looks like if you make a graph of the positions of the magnets at each step. Underneath each data point is a mini diagram showing the arrangement of magnets at that step. Why did I make this graph? Turn the page and you'll see what it's leading up to.

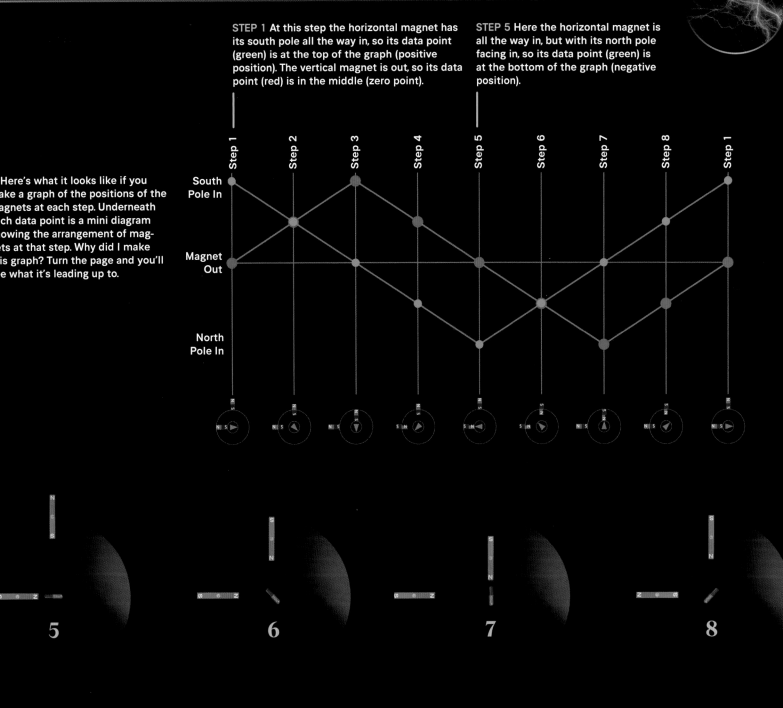

5

6

7

8

P 5 Here we are almost back to 1, with the horizontal magnet all way in and the vertical magnet out, this time the horizontal magnet ts north pole in. The rotor has ed 180 degrees from its position ep 1.

STEPS 6-8 Repeating the process, we can bring the rotor all the way around a complete circle and back to Step 1.

Finally, a Motor!

THE MODEL ON the previous page might not seem like much progress: we've just made a magnet rotate by moving other magnets around it. But remember that we can substitute a coil of wire for a permanent magnet, and then we can turn its magnetic field strength up and down, or even reverse it, by varying the voltage we apply to the coil.

We could make a jerky sort of motor by just turning the currents entirely on or off, but perhaps we can do better. Could we actually create a *smoothly* rotating magnetic field, with no discrete steps? If you were paying attention back in the steam engines chapter it should come as no surprise that the answer is yes, and we use sine and cosine waves to do it.

As we learned earlier, circular motion can be decomposed into two sine-wave motions in the x (left/right) and y (up/down) directions, with the two waves 90 degrees out of phase (shifted relative to each other). Because it's hard to visualize invisible magnetic fields, I've made a diagram of this arrangement of coils, with arrows that represent the strength of the magnetic field being generated by each coil, and a big pointer that shows the direction of the combined field in the middle. You get the direction of this pointer by adding together the arrows for the individual fields end to end. (This is called vector addition.)

What these plots show is that if you apply two voltages each following a sine wave pattern, one shifted by 90 degrees from the other, the magnet will spin. This is called a two-phase AC motor.

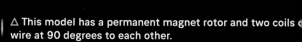

△ This model has a permanent magnet rotor and two coils of wire at 90 degrees to each other.

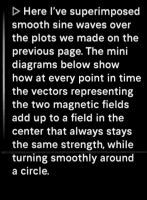

▷ Here I've superimposed smooth sine waves over the plots we made on the previous page. The mini diagrams below show how at every point in time the vectors representing the two magnetic fields add up to a field in the center that always stays the same strength, while turning smoothly around a circle.

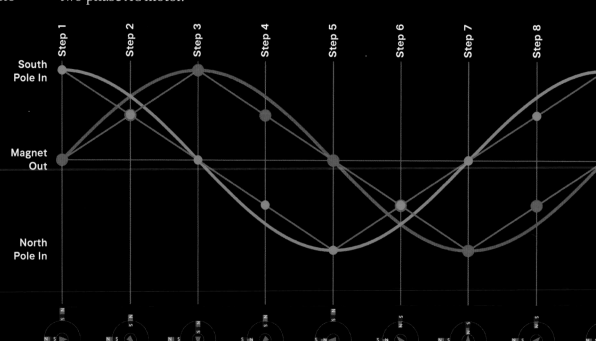

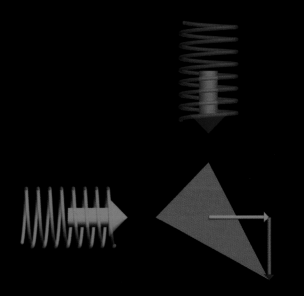

At **Step 1** in the previous example, the horizontal magnet was fully inserted, while the vertical magnet was pulled out. With coils instead of magnets this corresponds to full voltage applied to the horizontal coil and no voltage on the vertical coil. The pointer points right, because only the horizontal field is contributing to its direction.

At **Step 2** both magnets were partially inserted, which corresponds to sending a partial voltage to both of the coils. The direction of the pointer is found by adding the two arrows end to end. It doesn't matter which one you start with, you'll always get to the same endpoint when you string the arrows together.

When Nikola Tesla Wasn't Right

NIKOLA TESLA has a reputation as the world's most brilliant inventor. He's often talked about in connection with Thomas Edison, with whom he feuded for years about how electricity should be distributed: He believed in AC distribution, while Edison favored DC distribution. Tesla was right about many things, including that AC works better than DC in nearly every way. He was brilliant. He made many important discoveries in the early history of electricity and magnetism. But his electric motor was not the best one of its time.

Tesla designed an AC motor that worked as shown on the previous few pages, with two phases shifted 90 degrees. This two-phase motor works well enough, but it was never as good as what we're going to learn about next: three-phase motors. It turns out that taking the step from two to three phases, while confusing at first, has a couple of *huge* advantages.

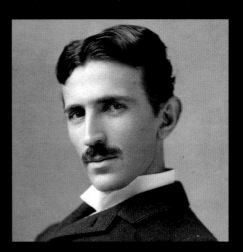

△ Nikola Tesla designed and built a two-phase electric motor in 1887. It worked, and it was important, but it wasn't the future of motors.

△ Mikhail Dolivo-Dobrovolsky may not have been as handsome as Tesla, but he designed a three-phase motor in 1888 and went on to create a whole ecosystem of three-phase generators, power lines, and motors. It is this system, not Tesla's, that took over the world.

The Real Prize:
A Three–Phase AC Motor

THE TWO-PHASE MOTOR we've been studying so far is good, but it's not great. Creating a circle out of sine and cosine waves is obvious once you've studied these curves. It's something any competent engineer would think of first (once they realized what problem needed to be solved). True greatness came by taking the next step and realizing that it was possible to do better by using three sine waves instead of two.

The motor on this page has three coils instead of two, and they are arranged evenly around the circle, with a spacing of 120 degrees between each. This arrangement means there are three magnetic fields to work with. Three sine waves of voltage, each 120 degrees out of phase with each other, together result in a smoothly rotating magnetic field in the middle.

Why is this better than getting the same result using just two coils and two sine waves? There is some advantage in the fact that the extra coils make the rotation of the field smoother. But the real advantage is that this arrangement allows for more power to be transmitted through less copper. As we will see shortly, a three-phase system can deliver *twice* as much power per pound of copper wire than a single-phase or two-phase system.

This increased efficiency is what makes three-phase power *such* a good idea that basically the entire global electric power grid is built around it. It's what all large generators create, what's sent through all large AC distribution lines, and what's delivered to all large industrial electric customers. The simpler, single-phase power available in most homes is derived from three-phase power somewhere pretty close to the end user, in a substation or on a power pole.

It's very hard to make a diagram that communicates exactly how the fields in a three-phase motor work, but I've done my best. I'm using three colors—red, green, and blue—to represent the three coils. Each one is fed a varying voltage shown by the red, green, and blue lines in the graph. And to represent the strength and direction of the magnetic fields, I'm using individual red, green, and blue arrows to represent the contributions of the individual coils, and a big gray arrow to point out the direction of the combined field created by all three working together. As in the last section, the magnetic fields of each coil add together by vector addition, which takes into account both their strength and direction.

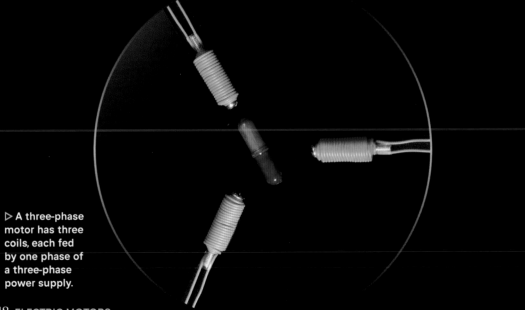

▷ A three-phase motor has three coils, each fed by one phase of a three-phase power supply.

▷ In this diagram of three-phase power, the vertical axis is voltage and the horizontal axis is time. The voltage is typically 208 volts, and each cycle is either $1/50$ or $1/60$ of a second, depending on the country you're in.

The pattern of three sine waves is special because, even though all three of the fields are constantly varying in strength, the *total* strength of the field is *exactly the same at every position*. And it's revolving at a smooth, constant rate. Only sine waves, arranged this way, will give you a field that both rotates at a constant rate, and stays at a constant strength at all times.

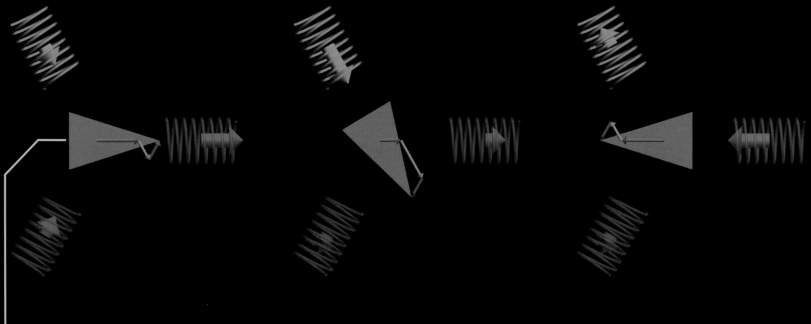

STEP 1. At this point in the cycle the red phase is at its maximum voltage, so the red magnet is pulling the field strongly to the right. The green and blue coils are partially on in the opposite direction (with their fields pointing in rather than out). But notice that green and blue work together to help the overall field point toward the red coil more strongly.

STEP 2. A little bit later the situation is more complicated. None of the voltages are at full strength, but when you add them up the result is a field pointing midway between the red and blue coils (notice the three colored vectors summing to give the field indicated by the gray triangle).

STEP 3. Halfway through a revolution of the motor, the red field is again at full strength, but this time it's pointing in the opposite direction, pushing inward. The green and blue fields again add up to help the field along in the same direction.

1   2   3

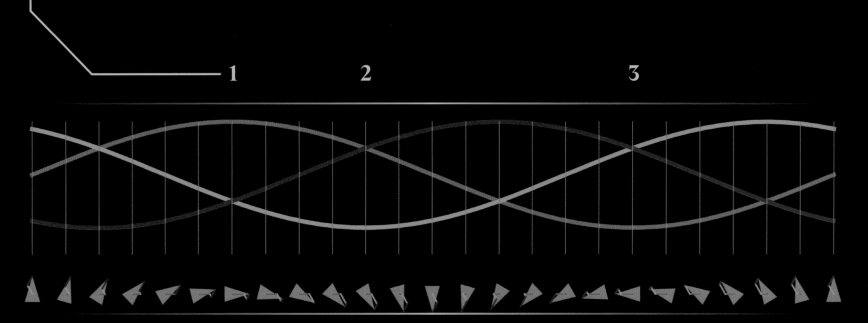

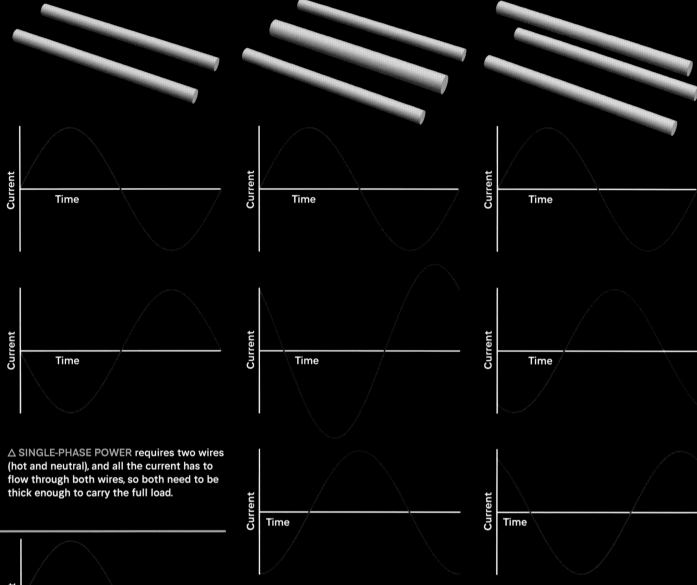

▷ Why is three-phase power uniquely the best solution? Because three-phase motors and generators are more efficient, and three-phase power lines can distribute more power with less going to waste in the wires.

△ SINGLE-PHASE POWER requires two wires (hot and neutral), and all the current has to flow through both wires, so both need to be thick enough to carry the full load.

SPLIT PHASE

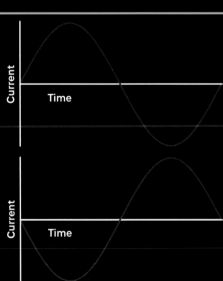

▷ It's important to note that two-phase power is *not* the same thing as the "split-phase" power commonly available in the US and other countries. Split-phase power is the same thing as single-phase power, with a tap in the middle of the transformer that allows for two sides of the same phase.

△ TWO-PHASE POWER requires either four wires (two for each phase), or three wires (phase 1, phase 2, and neutral). If you use three wires, during part of the cycle nearly twice as much current flows through the neutral as through the two phases. The two hot wires need to be thick enough to carry their full load, and the neutral needs to be almost twice as thick, meaning you're using only slightly less copper than for two separate single-phase lines. So the power-per-pound-of-copper is little better than for single-phase power.

△ THREE-PHASE POWER also requires three wires (phase 1, phase 2, and phase 3), but there is no neutral wire. All three wires can be the same thickness, and all of them carry the same amount of current. For a given thickness of wire, and a given amount of power delivered to the destination, a three-phase system wastes significantly less power in the wires as would a single-phase system using the same amount of copper (or aluminum in the case of large diameter wires).

Fun Application of Three-Phase Motors

THREE-PHASE MOTORS with permanent magnet rotors, as explained in the previous section, are among the most compact, powerful electric motors around. On a power-to-weight basis they can't be beat. One potential disadvantage of these motors is that they can only operate exactly in sync with the frequency of the AC power being sent to them. If they are running off power from the electric grid, they can only run at one fixed speed.

You can sidestep this problem by creating your own three-phase power locally from a battery (for example, in a drone or electric car). Using modern high-power semiconductor chips, it's possible to make a very small control box that creates hundreds or even thousands of watts of three-phase electric current, at any frequency you like. This lets you spin the motor not at a fixed speed, but at a very precisely controlled speed. Unlike DC motors, where you vary the speed by varying the power, with these motors you vary the speed by varying the frequency. This lets them run at full torque (turning force) at any speed, which in turn allows them to change speed almost instantly—important when you need precise control over the thing the motor is turning.

Modern quadcopter drones, for example, are possible because of four factors coming together: First,

super-strong neodymium iron boron magnets are used for the rotors, which make the motors strong and light. Second, high-power semiconductor chips allow the three-phase power to be created efficiently. Third, high-speed, low-power microprocessors implement the complicated, real-time feedback loops that keep the things stable in the air. Fourth and finally, lightweight lithium polymer batteries allow enough

power to be stored for them to run for a reasonable length of time.

When a drone is flying in turbulent air, the individual propellers must respond in milliseconds to changing conditions. Traditional DC motors wouldn't be powerful enough, but they also wouldn't be able to speed up and *slow down* fast enough. Three-phase permanent magnet motors are the perfect solution.

You can get an idea of how powerful these motors are from the thickness of the wires coming out of them. Even though the motors are less than 2 inches across, these wires are thick enough to carry tens of amperes of current.

▽ Drones typically use "outrunner" motors, where the stationary coils are on the inside. Permanent magnets are mounted to the outer shell, which spins with the shaft. In other words, what looks like the motor is actually the rotor, the part that spins. The only visible part of the motor that *doesn't* spin is the cap on the bottom, which has threaded screw holes for mounting.

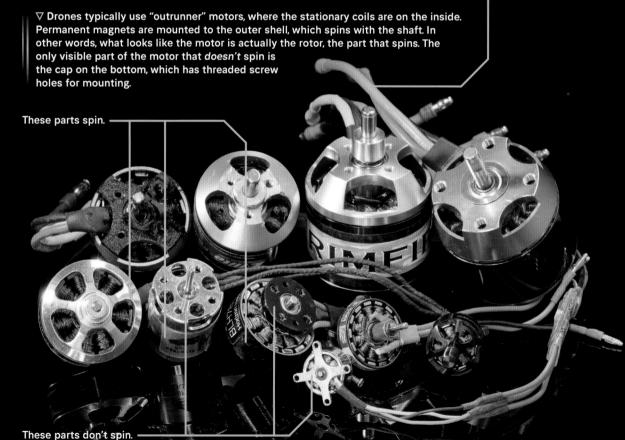

These parts spin.

These parts don't spin.

▷ Taken apart, you can see the coils in the core of this fine example of a drone outrunner motor. On the inside, which is stationary and bolted to the frame of the drone, there are twelve coils. Lining the inside of the outer case—the part that spins—are fourteen powerful neodymium-iron-boron magnets. The configuration is more complicated than in my model, but the principle is the same: a rotating magnetic field created by the twelve coils operating in groups of four each, driven by three-phase power, pulls permanent magnets around with it. The extreme power of this motor comes from the combination of the large currents flowing through the coils, which create powerful magnetic fields, and the high strength of the permanent magnets.

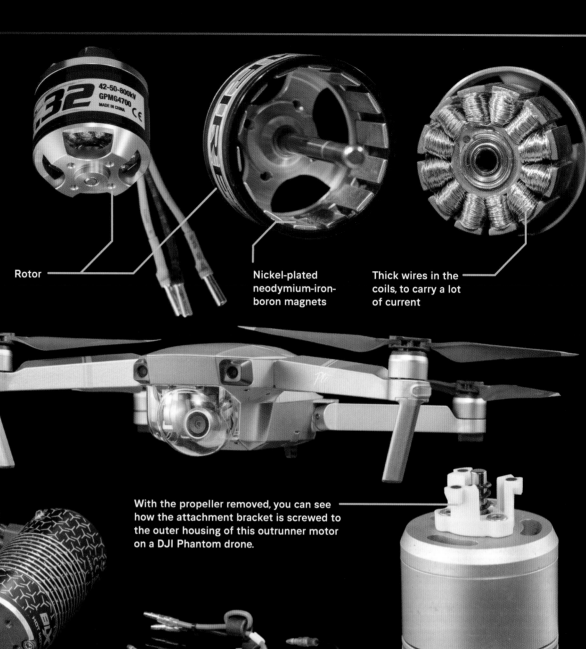

Rotor

Nickel-plated neodymium-iron-boron magnets

Thick wires in the coils, to carry a lot of current

Outrunner motor on a DJI Mavic drone

With the propeller removed, you can see how the attachment bracket is screwed to the outer housing of this outrunner motor on a DJI Phantom drone.

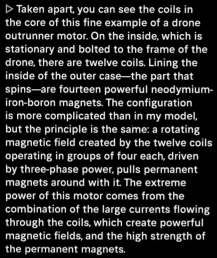

△ High-performance remote control models that *don't* fly more typically use in-runner three-phase permanent magnet motors. They are functionally identical to the outrunners we just saw, except inside out—in other words, normal, with the rotor on the inside.

◁ At the budget end of the model motor spectrum we have these conventional brushed DC motors. Not only are the motors themselves cheaper, they also don't require a fancy electronic speed controller to generate three-phase power for them. Their speed is controlled by varying the voltage, which is a much sloppier way of doing it, and not suitable for the precision required in a drone.

▷ Neodymium magnets are scary strong. The outrunner motor shown left gave me a nasty pinch—with blood and everything—when I put it back together. The attraction of the magnets to the iron core was so strong that I lost my grip on the two halves and they slammed into each other with considerable force. Taken a step further, two large neodymium magnets let loose near each other will make mincemeat of fingers, or entire hands. The force required to separate them can be thousands of pounds for magnets just an inch (2.5 cm) across.

△ Drones became possible because of electric motors with a high power-to-weight ratio. This remarkable thing was built to find out whether a human being has the necessary power-to-weight ratio to be the engine in an airplane. The answer is yes, but (a) you need a really strong person, and (b) you need a really light airplane. This one, the *Gossamer Condor*, piloted on a record-breaking flight by competitive cyclist Bryan Allen, is basically made of plastic and Saran wrap, and it weighs just *70 pounds (32 kg)*—half the weight of the pilot! As impressive as this achievement is, note that this is a winged airplane, which

△ Although it's fairly ridiculous to think it's possible, a human-powered quadcopter has actually been built, and flown for 64 seconds to an altitude of 11 feet (3.3 m). So, it's *possible*. . . . But compare this 64-second record to the fact that the *Gossamer Albatross* (successor to the *Gossamer Condor*) flew all the way across the English Channel under human power. Airplanes are a *lot* more efficient than helicopters!

A Tale of Two Cars

◁ This electric car, a Flocken Elektrowagen, was made in 1888, before there were practical gasoline-powered cars.

△ The gasoline-engine Ford Model T was not introduced until twenty years later, in 1908.

▷ Everything new is also old. Hub motors may be the latest thing in electric cars, but the first gasoline-electric hybrid car, this Lohner-Porsche from around 1900, had hub motors. Ferdinand Porsche actually started his famous company making electric cars!

Hub motors are integrated into the wheels.

ELECTRIC CARS may soon replace gas-powered cars, but the electric motor actually predates the internal combustion engine by several decades. There were even electric cars before there were gasoline-engine cars! Of course, with the primitive battery technology of the day, their range was incredibly small; basically they could get across town and maybe back.

But despite this drawback, for several decades, from the late 1800s to the mid-1910s, electric cars dominated the road, even winning land speed records. That's likely because the main competition at the time were steam-powered cars that took half an hour to warm up in the morning, and gasoline-powered cars before the invention of the muffler and the starter motor. Electric cars of that

time were successful not because they were especially good cars, but because the alternatives were really terrible.

Once large-scale production of good gasoline engines began (notably by Henry Ford), electric cars were washed away by a flood of affordable cars able to drive long distances on a tank of cheap, portable gasoline. Today the tide is turning again, now that batteries and electric motors have gotten *much* better. The advantages of electric cars go well beyond the environmental benefits. Once the battery problem is completely solved—and we're very close—no one will want anything else.

One huge advantage to electric motors is that they are able to operate efficiently at full torque (rotational force) right down to zero

rpm. That means electric cars can leap into motion instantly from a dead start. Gasoline engines work well only within a fairly narrow range of rotational speeds, and have hugely reduced torque at low speed. After the engine, the second-largest—and second-most-expensive—part of a gasoline-powered car is the transmission, whose job is to keep the engine spinning near its optimal speed even while the vehicle is moving anywhere from very slow to very fast. Electric cars simply don't need a transmission.

Gasoline cars also operate with the help of a *differential*, which allows a single engine to power two wheels at different speeds. When you're going around a corner, for example, the outer wheel needs to go faster than the inner wheel. While many electric cars are also built with a differential, the trend is toward simply putting in two electric motors—or even four, one per wheel. Then you can eliminate *all* the gears between the engines and wheels.

In theory, electric cars could even be equipped with hub motors. This is an electric motor mounted literally inside the wheel. The obvious advantage of this arrangement is that exactly zero room is taken up within the car. However, there's a less obvious disadvantage to a hub motor—"unsprung weight." Car designers go out of their way to reduce the amount of weight that is not separated from the ground by springs and shock absorbers, because it causes handling problems when the wheels go over bumps. Hub motors move a lot of weight directly into the wheels, where it is unsprung. It will be interesting to see how this plays out over the coming years. Will hub motors turn out to be the best solution, or a great solution only until we realize it's actually terrible? At least with electric motors we have the opportunity to find out.

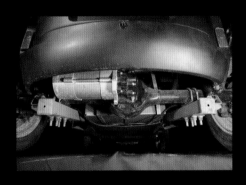

△ As of this writing I do not have a serious electric car, but I do have this overgrown toy! The streets of Beijing are filled with these cute little buggies in every shape and style. They don't go very fast or very far, but they are incredibly efficient and fun to drive. My plan was to drive it the mile or so back and forth to my studio every day. There's no street on my route with a speed limit over 30 mph, and the battery could easily take me back and forth several times, so what's the point of using a car that could go 100 mph for 400 miles? Sadly, the city council resisted all my attempts to get them to change the city ordinances on this type of car, even after I showed them pictures of how cute it is. So now it's just a tax-deductible research object to put in my book, and stare at longingly.

△ Seen from below, the entire drivetrain of this micro-car is ridiculously simple. It's just an electric motor directly coupled to a differential, all mounted on a single rigid rear axle. The differential could be eliminated by having two motors, but motors are currently more expensive than differentials, and this car is optimized for cheapness! In a minimalist gasoline car with similar performance, the drivetrain would be many times larger and weigh hundreds of pounds. This one's is no bigger than a bread box, not counting the axle, and I could lift it with one hand. (The "gas tank" consists of six ordinary lead-acid car batteries housed under the driver's seat.)

◁ If you think my minicar looks silly, well, get a load of this one! On the other hand, it's street-legal, which is more than I can say about mine. On the third hand, it costs five times as much. Anyway, the point is that we are now in a time of tremendous innovation in the electric car world. Freed from the mechanical limitations imposed by internal combustion engines, people are trying out all kinds of new ideas. In a few decades many of them will seem ludicrous, and a few of them, maybe including this one, will be everywhere. (No, the picture is not cut in half. The car is. But I do want to add that other car models from the same company are really quite attractive.)

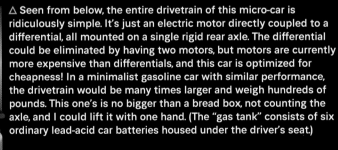

An Alliance of Worlds

Revolving Field Alternating-Current Generator

High Speed Engine

Exciter Belt Drive

Direct-Current Exciter

The high-speed steam engine spins a shaft.

The electric generator converts the energy of the spinning shaft into an electric current.

△ This image from around 1915 shows a steam engine connected to a generator to supply electric power to a factory.

ELECTRIC MOTORS have a long history of working literally side by side with gasoline, diesel, and steam engines. Electric motors assume a supply of electricity, which has to come from somewhere. In an ideal world that might be from solar cells, windmills, a nuclear reactor, a tide-powered generator, a geothermal power plant, or some other nonpolluting source. But in practice the energy needed to run an electric motor often comes from a generator being spun by an engine burning some kind of fuel. (Generators are explained in more detail in a few pages: basically they are devices that take some form of energy—for example, supplied by a spinning shaft—and turn it into electrical energy.)

Because electric motors are so clean and quiet, people are happy to have them running nearby, even if that means having a bigger, noisier fuel-burning engine somewhere farther away. Imagine if your kitchen mixer was powered directly by a noisy, smelly, carbon monoxide–generating internal combustion engine? It's much more pleasant to keep the fuel-burning engine as far away as possible—maybe out in the garden shed, or maybe in another state—running a generator.

▷ Here's a gasoline engine running an electric generator, which is connected by a cord to an electric motor, which is spinning the blades of a lawn mower. See how convenient that is? OK, no, this is a joke, because the gas motor would be perfectly fine turning the mower blades directly, cutting in half the number of kids required to mow the lawn. But this is in fact *exactly* how many trains are built (minus the kids).

Generator
Gas engine

Mower with electric motor

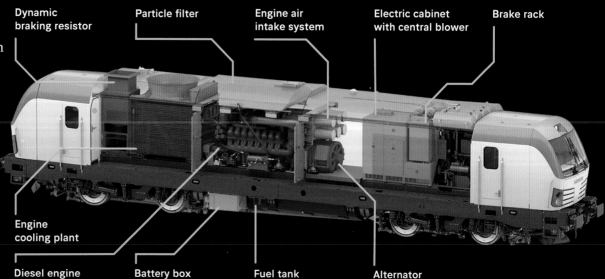

Dynamic braking resistor

Particle filter

Engine air intake system

Electric cabinet with central blower

Brake rack

Engine cooling plant

Diesel engine

Battery box

Fuel tank

Alternator

▽ These photos from a book published in 1917 show an installation at a mining company that used a steam engine to run a generator to run an electric motor to run the hoist used to pull ore and miners up from the depths. They could have connected the steam engine directly to the hoist, but there's a good reason why the more complicated electric setup made sense. An electric motor can work as a generator too, if you spin its shaft. The mine setup had a 50-ton (45 metric ton) flywheel that stored a *lot* of energy by spinning very fast. When the hoist was going down the mine shaft, its descending weight spun the hoist motor, turning it into a generator, which supplied power to a motor connected to the flywheel. The potential energy released by the descending hoist sped up the flywheel. When it was time to raise the hoist back up, the motor connected to the flywheel turned into a generator and power was drawn from the flywheel to run the hoist motor. The steam engine only needed to supply a relatively modest amount of power to make up for friction losses, and to compensate for the fact that typically more stuff comes up from a mine than goes down.

Switchboard — Torque Motor — 1400 h.p. A.C. Motor — 1500 kw. Direct-Current Generator — Slip Regulator — Casing Containing 50-Ton Fly Wheel

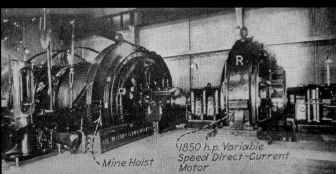

Mine Hoist — 1850 h.p. Variable Speed Direct-Current Motor

◁ A typical diesel locomotive is set up *exactly* like the ridiculous example of the lawn mower at left. An internal combustion engine drives a generator, which drives an electric motor, all within the confines of the locomotive. It's not ridiculous in this case, because while internal combustion engines are fine for spinning lawn mower blades, they are terrible for getting long, incredibly heavy trains to move from a standstill. They can't deliver high torque at low speeds, whereas electric motors actually generate *more* torque when they are going slower. The generator, motor, and the associated control and power management systems replace the function of what would otherwise have to be a huge mechanical transmission.

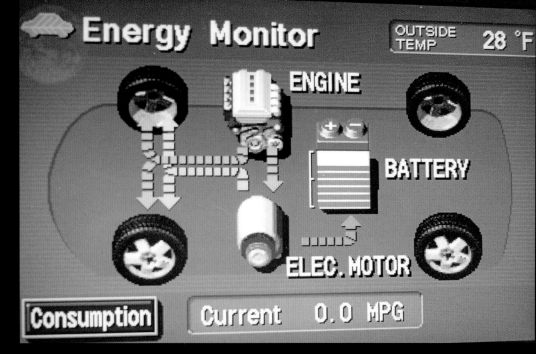

Energy Monitor

OUTSIDE TEMP 28 °F

ENGINE

BATTERY

ELEC. MOTOR

Consumption

Current 0.0 MPG

△ Using drive motors as generators to recapture energy is common in electric cars, where it is known as regenerative braking. Instead of storing the recovered energy in a flywheel, it's used to recharge the battery. Can you make a perpetual motion machine out of this? No, because on level ground you'll never get as much energy back from braking as you need to put in to get the car up to speed. But if you start at the top of a mountain and go downhill for miles and miles, you might well end up with more charge in the battery than you started with. Just don't expect to be able to get back up the hill on that amount of charge.

A 220-volt split-phase AC generator

Four-stroke internal combustion engine

△ Back when I was more high-strung, I installed this generator on my farm. It's got a 55-horsepower car-sized engine that runs on propane, and can supply as much power as my connection to the electric grid. In retrospect, this was overkill.

Motors Without Magnets

A MOTOR NEEDS two magnetic fields that push against each other, but it doesn't necessarily need any magnets. Both magnetic fields can come from coils of wire, one on a stationary frame and the other connected to the rotating shaft of the motor. Using two electromagnets like this has a number of advantages, but one huge potential disadvantage: How do you feed current to the spinning electromagnet? Without current, your electromagnet is just a coil of wire.

One obvious solution is to use sliding contacts, as in the DC motors earlier in this chapter. A type of motor called an AC/DC universal motor (no relation to the rock band) is built exactly the same way as a DC brush motor, except with coils instead of permanent magnets on the outside. Consider what happens if we run this motor on DC current from a battery. The coils on the outside act just like permanent magnets, and the motor runs normally. Now imagine we reverse the battery connection: This flips the direction of current in the rotating coils on the inside, which reverses the direction of the motor. But it also flips the polarity of the electromagnets on the outside, which *also* reverses the direction. Two reversals mean we're back to going in the same direction.

Operated on DC from a battery, the motor turns in the same direction regardless of which way you connect it. That means you can reverse the polarity as many times as you like, and it will keep spinning in the same direction. In other words, this type of motor works the same on AC as it does on DC! These motors are powerful, cheap, and noisy; you typically find them in vacuum cleaners and other loud appliances.

It would be really cool if it were possible to make a motor that has no permanent magnets, and *also* doesn't have any sliding electrical connections. Is that possible? How would you get electric current into the coils in the rotor without some kind of sliding connection? Well, it *is* possible, but we're going to have to take an extended detour through generators and transformers before we can understand how.

Commutator identical to one in a DC brush motor

Rotating coils identical to those in a DC brush motor

Electromagnet coils that take the place of permanent magnets

A Brief Interlude to Discuss Generators and Transformers

IT'S IMPOSSIBLE TO continue our discussion of motors without first talking a bit about generators and transformers. These devices are very closely related to motors, and it was only by understanding the principles behind them that people like Nikola Tesla were able to invent the profoundly clever types of motors we're going to learn about next.

So far in our discussion of motors, we've concentrated on the fact that current flowing in a coil of wire creates a magnetic field. Now we're going to look at the complementary fact that if you move a coil of wire through a magnetic field, a current is created (induced) in the wire.

A battery-free "shake" flashlight is an example of the simplest possible version of a magnetic field moving through a coil of wire. When you shake the flashlight, a magnet literally slides back and forth through a coil inside. When the magnet passes through the coil it creates a current in the coil, which is sent to the bulb, causing it to light up.

Now, what do we know about currents in coils? They create magnetic fields! So when the magnet is moving through the coil, we actually have two magnetic fields—one from the permanent magnet, and a second from the current that was created in the coil. It works out that the field from the current in the coil will always point in the opposite direction from the field of the magnet, so they will push against each other. That means it takes work to get the magnet to move through the coil, and it is this work that supplies the energy that lights the bulb.

In other words, when the magnet moves past the coil of wire, they push against each other with invisible magnetic force, just as two magnets will push against each other. But in the case of the magnet and the coil, the force is felt *only when the magnet is moving.* That's because the magnetic field coming from the coil only exists while the magnet is moving across it. The field in the coil is coming from the current in the coil, which in turn is coming from the *movement* of the magnet. When the magnet stops, the current stops, the field from the coil goes away, and the force disappears. In addition, the force created from the movement of the magnet will always try to slow down the magnet, regardless of what direction it's moving.

This circular interplay between moving fields, induced currents, and secondary fields from those currents is central to the way that generators and transformers work, and it's the basis for the new kind of motor we're going to talk about after our detour.

Small magnets at each end of the barrel are arranged so they repel the big sliding magnet, bouncing it smoothly back and forth as you shake the flashlight.

Heavy, strong magnet that is free to move back and forth

Coil with many turns of fine wire

The electric current is supposed to be saved up in this battery, so you don't have to shake it *all the time.* Except this one is so cheap the battery doesn't work, making this the ultimate horror movie flashlight: As you frantically shake it, the monster is closer each time you get a brief flash of light in its direction! Aaaahhhh!

Generators

▷ You can induce currents in solid pieces of metal, not just in coils of wire. For example, here is a very strong permanent magnet being moved back and forth close to a thick copper plate. When you move it fast, it feels like you're moving it through molasses. The current induced in the copper plate by the movement of the magnet pushes back at you. It's really quite an odd sensation. But if you move it slowly, the induced current is weak, and there is little resistance to movement. (The fact that the slow blade penetrates the shield, as we learn in *Dune*, means that the shield is an electromagnetic induction field. Or something like that.)

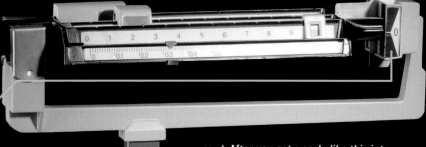

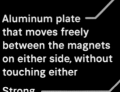

Aluminum plate that moves freely between the magnets on either side, without touching either

Strong magnets on both sides

MB-311
CAPACITY 311g
SENSITIVITY 0.01g

QUADRUPLE
BEAM BALANCE
MADE IN CHINA

△ After you get a scale like this into near-perfect balance, it tends to oscillate up and down for a long time before settling on a reading. Everything that makes the balance more accurate—sharp knife edges and ultra-hard materials in the pivot points—also reduces friction, which means it will oscillate for longer. But the last thing you want to do is add friction to make it stop, because that would prevent it from coming to rest exactly where it should. Instead, these balances commonly use a magnetic damper: two strong magnets on either side of a metal plate attached to the end of the balance arm. When the balance is moving up and down rapidly, induced currents in the plate work to slow it down. When it's moving slowly near the end, those currents reduce to zero, and thus do not influence the final stopping point.

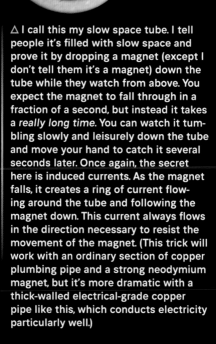

△ I call this my slow space tube. I tell people it's filled with slow space and prove it by dropping a magnet (except I don't tell them it's a magnet) down the tube while they watch from above. You expect the magnet to fall through in a fraction of a second, but instead it takes a *really long time*. You can watch it tumbling slowly and leisurely down the tube and move your hand to catch it several seconds later. Once again, the secret here is induced currents. As the magnet falls, it creates a ring of current flowing around the tube and following the magnet down. This current always flows in the direction necessary to resist the movement of the magnet. (This trick will work with an ordinary section of copper plumbing pipe and a strong neodymium magnet, but it's more dramatic with a thick-walled electrical-grade copper pipe like this, which conducts electricity particularly well.)

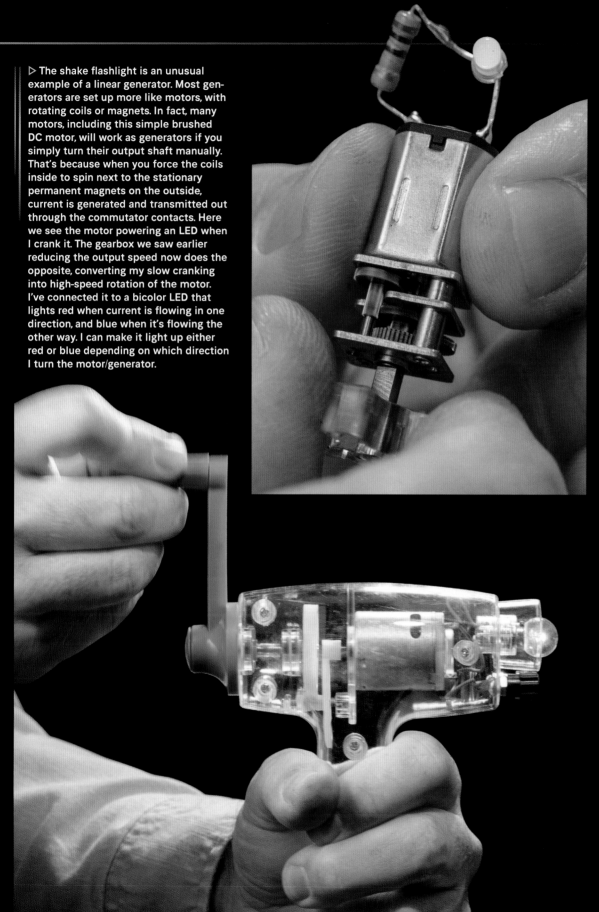

▷ The shake flashlight is an unusual example of a linear generator. Most generators are set up more like motors, with rotating coils or magnets. In fact, many motors, including this simple brushed DC motor, will work as generators if you simply turn their output shaft manually. That's because when you force the coils inside to spin next to the stationary permanent magnets on the outside, current is generated and transmitted out through the commutator contacts. Here we see the motor powering an LED when I crank it. The gearbox we saw earlier reducing the output speed now does the opposite, converting my slow cranking into high-speed rotation of the motor. I've connected it to a bicolor LED that lights red when current is flowing in one direction, and blue when it's flowing the other way. I can make it light up either red or blue depending on which direction I turn the motor/generator.

◁ The exact same idea—a crank that turns an electric motor/generator connected to a light bulb—is available as a commercially produced educational demonstration.

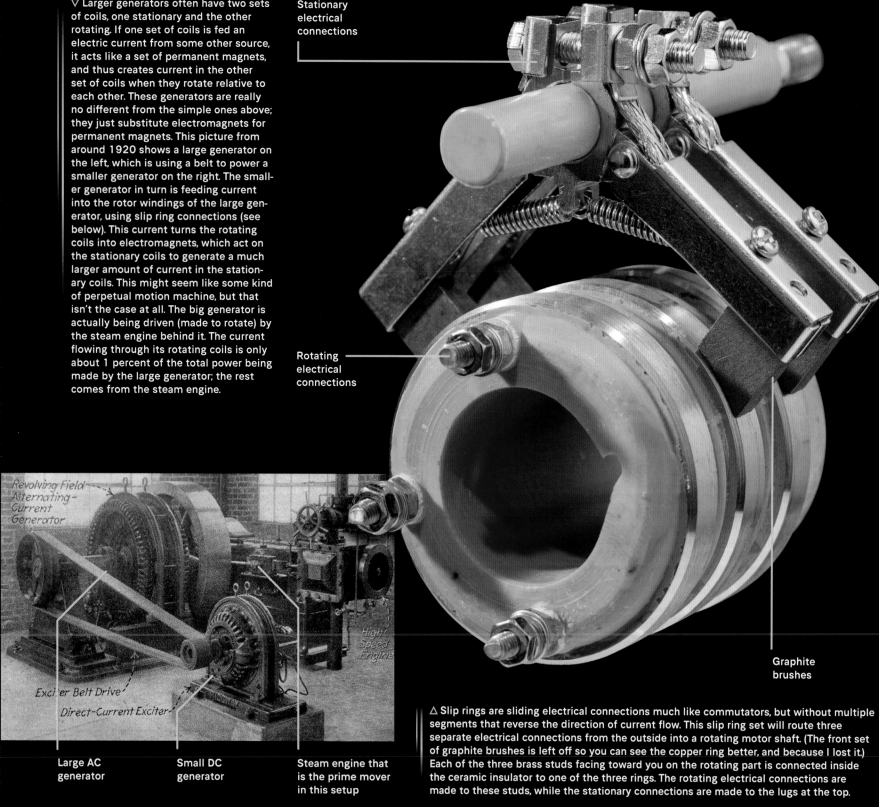

∇ Larger generators often have two sets of coils, one stationary and the other rotating. If one set of coils is fed an electric current from some other source, it acts like a set of permanent magnets, and thus creates current in the other set of coils when they rotate relative to each other. These generators are really no different from the simple ones above; they just substitute electromagnets for permanent magnets. This picture from around 1920 shows a large generator on the left, which is using a belt to power a smaller generator on the right. The smaller generator in turn is feeding current into the rotor windings of the large generator, using slip ring connections (see below). This current turns the rotating coils into electromagnets, which act on the stationary coils to generate a much larger amount of current in the stationary coils. This might seem like some kind of perpetual motion machine, but that isn't the case at all. The big generator is actually being driven (made to rotate) by the steam engine behind it. The current flowing through its rotating coils is only about 1 percent of the total power being made by the large generator; the rest comes from the steam engine.

Stationary electrical connections

Rotating electrical connections

Graphite brushes

Revolving Field Alternating-Current Generator

Exciter Belt Drive

Direct-Current Exciter

High Speed Engine

Large AC generator

Small DC generator

Steam engine that is the prime mover in this setup

△ Slip rings are sliding electrical connections much like commutators, but without multiple segments that reverse the direction of current flow. This slip ring set will route three separate electrical connections from the outside into a rotating motor shaft. (The front set of graphite brushes is left off so you can see the copper ring better, and because I lost it.) Each of the three brass studs facing toward you on the rotating part is connected inside the ceramic insulator to one of the three rings. The rotating electrical connections are made to these studs, while the stationary connections are made to the lugs at the top.

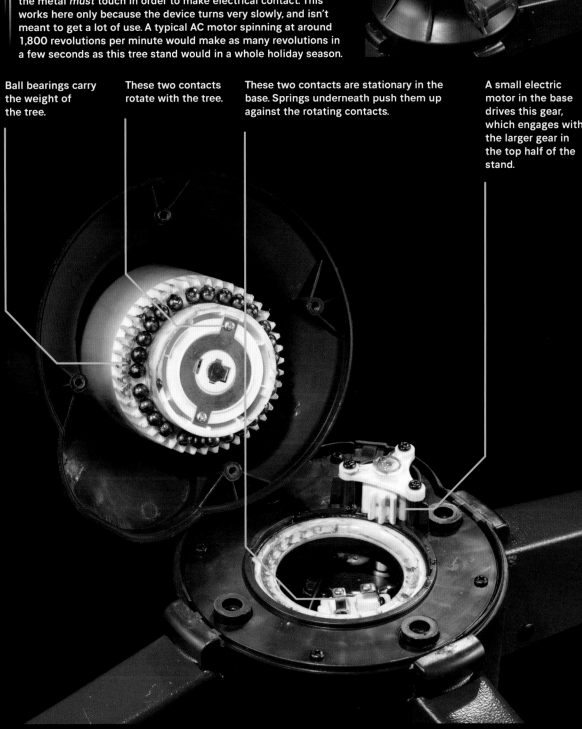

▷ This rotating Christmas tree stand has a copper-to-copper slip ring connection to feed power to the lights on the tree as it turns. Metal sliding directly against metal generates a lot of friction, so in mechanical devices oil is normally used to prevent the metal surfaces from actually touching each other. But in this situation the metal *must* touch in order to make electrical contact. This works here only because the device turns very slowly, and isn't meant to get a lot of use. A typical AC motor spinning at around 1,800 revolutions per minute would make as many revolutions in a few seconds as this tree stand would in a whole holiday season.

Ball bearings carry the weight of the tree.

These two contacts rotate with the tree.

These two contacts are stationary in the base. Springs underneath push them up against the rotating contacts.

A small electric motor in the base drives this gear, which engages with the larger gear in the top half of the stand.

△ This is an odd artifact from the past: a rotating contact, similar to a slip ring, that uses a pool of liquid mercury metal to make the electrical connection, instead of a graphite brush. In theory this is a great idea. In practice … it's mercury. One notable application was in the Leslie loudspeakers used by some of the most famous musicians of the 1960s. They loved the unique sound created by the fact that these speakers have a treble driver (a small speaker for high-frequency sounds) that spins inside the cabinet. This requires a rotating electrical connection, and if you tried to use a graphite brush slip ring, you would hear nothing but electrical noise as the contacts bounced and scraped along. Liquid mercury makes for a completely noise-free, uninterrupted connection. But … it's mercury. The company that made these is out of business, but they are readily available on eBay for repair and restoration work.

Generators That Run on Air and Water

MOST GENERATORS ARE of the fire nation: they are driven by engines running on some kind of burning fuel. But more and more generators are being made to run on wind and flowing water. As long as you can get a shaft to rotate with a reasonable amount of force, you can hook a generator up to it and turn that rotation into electric power.

A wind farm was recently installed all around my farm compound in Illinois. The nearest windmill is only about a mile away. A windmill is basically a generator on a stick. Some smaller windmills use permanent magnet generators, much like the toy example we saw, but larger ones use rotating electromagnets fed through slip rings like the ones we saw on the previous page. This design can be scaled up better, and it also allows for the frequency of the output AC current to the power grid to be matched very accurately, even as the speed of the windmill varies randomly. This is done by running a complementary frequency of AC through the rotor windings. As the wind speed decreases, the frequency of the rotor winding current is increased, keeping the generated output frequency exactly in sync with the power grid.

▷ Windmills are popping up everywhere.

△ The box on the end of the stick houses the generator, a massive gearbox, steering motors, control electronics, and a cooling system (note the radiator sticking up near the back end). In case you're wondering, yes, in order to get this photograph, I flew my drone between the blades, but don't worry, it was a windless day and the farm was not yet online, so there was no electrical activity in these units. They are so tall that to take this picture I had to fly right up to the legally imposed maximum altitude of 400 feet (120 m), beyond which amateur drones refuse to fly. (According to the company installing them, these are the tallest type of windmill installed anywhere in the US.)

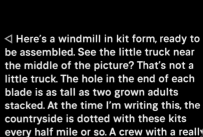

◁ Here's a windmill in kit form, ready to be assembled. See the little truck near the middle of the picture? That's not a little truck. The hole in the end of each blade is as tall as two grown adults stacked. At the time I'm writing this, the countryside is dotted with these kits every half mile or so. A crew with a really tall crane is moving systematically from one to the next putting them together.

◁ Water turbine generators are just like windmills, except the blades are pushed by water instead of air. They also typically use slip ring generators, for the same reason windmills do. And these generators can also be run as motors. Running a windmill in reverse to create breeze would be possible, but silly. Running a water turbine as a pump, on the other hand, can be very useful. This is a picture of the storage lake at the Seneca Pumped Storage Generating Station in Pennsylvania. Its water turbines can run either as generators powered by water flowing down from the lake, or as motors pumping water back up into the lake. The idea is to run them as generators when there is a high demand for electricity during the day, and then at night, when the dam and other power plants in the area can make more power than people need, run them as pumps to move the water back up to the high lake, ready to be let down again when electricity is needed. It's like a giant, complicated rechargeable battery.

Transformers

IT MIGHT SEEM that we are straying even further afield by going from generators to transformers, but trust me, this is all going to tie back together with motors in the next section.

We've already learned that a permanent magnet moving through a coil of wire can induce a current in that coil. But, of course, it's not the physical magnet that has this effect, but rather the magnetic field around that magnet. And magnetic fields can come not only from permanent magnets, but also from other coils of wire.

Transformers are basically two coils of wire pointed at each other, with alternating current fed into one of them (called the primary winding) and drawn off from the other (called the secondary winding). The current being forced through the primary coil creates an ever-changing magnetic field that crosses through the secondary coil, resulting in an alternating current flowing through that secondary coil.

Transformers get their name from the fact that the most useful thing they can do is change, or transform, one voltage into another.

▷ Earlier, we discussed the need for a high-voltage spark to ignite the fuel-air mixture in an internal combustion engine, and that an "ignition coil" creates this spark. This is a variation on the neon transformer below. The ignition coil has a primary winding with a small number of turns of thick wire, and wrapped around that a secondary winding with a huge number of turns of very thin wire. Instead of being fed a smoothly varying AC current, an unvarying 12-volt DC from the car's battery is run through the primary coil. Because the voltage isn't varying, the magnetic field is constant, and no current is induced in the secondary coil. But when it's time to make a spark, the voltage on the primary coil is suddenly cut off, causing the magnetic field to collapse rapidly. This creates a huge voltage spike through the many turns in the secondary coil. In effect the energy to make the spark is stored up in the magnetic field while the voltage is present, and then routed through the spark plug when the field collapses. You might think there would be a second spark when the voltage is reconnected—and there is some voltage induced when the field is built up—but the buildup is much slower than the collapse. This process happens many times per second.

▷ In this neat little neon transformer, 120-volt AC is applied to the primary coil on the right. When the voltage is at the peak of its cycle, a strong magnetic field is generated and transmitted through the iron core to the secondary coil on the left. A fraction of a second later when the input voltage passes through zero, the magnetic field collapses to nothing. Then it grows again in the opposite direction as the voltage reaches its negative peak. And so on.

The pulsing magnetic field from the primary coil pushes and pulls through the secondary coil, pumping the electrons in the wire back and forth. If the secondary coil has more loops of wire than the primary, the output voltage will be higher (for example, if there are twice as many turns in the secondary coil, the output voltage will be twice as high as the input). This transformer has thirty times as many windings in its secondary coil, so it creates a very high output voltage. (The secondary is smaller even though it has more turns because the wires inside are much thinner than

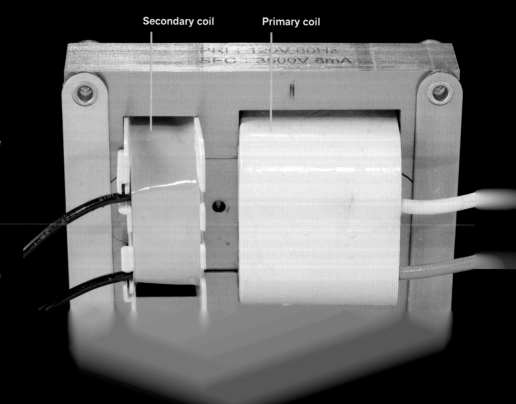

Secondary coil   Primary coil

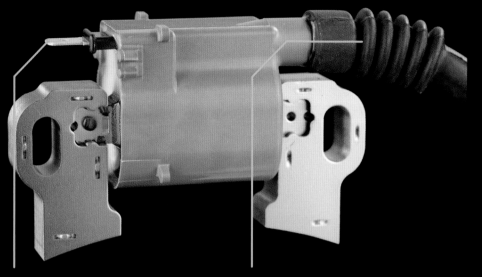

This "coil pack" serves a single spark plug.

A low-voltage signal from the control module tells the coil when to fire.

Twelve-volt DC from the battery goes in here.

This looks like a thick wire, but actually it's a thin wire with very thick insulation, necessary because of the high voltage it's carrying.

Spark plug (which would normally be screwed into the engine block)

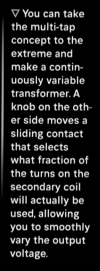

▷ Newer cars often have individual spark coils for each cylinder, driven by a solid-state engine management module. This eliminates the mechanical contacts in the distributor, and the high-voltage spark plug wires connecting them to the spark plugs.

▽ You can take the multi-tap concept to the extreme and make a continuously variable transformer. A knob on the other side moves a sliding contact that selects what fraction of the turns on the secondary coil will actually be used, allowing you to smoothly vary the output voltage.

▷ Most transformers have their primary and secondary coils wound together, on top of each other, so you can't see them as separate coils. It just looks like one block with many wires coming out of it. It's also common for there to be several "taps"—separate connections at different points—along one or both of the coils. This one has three taps in the primary (input) coil, so you can connect it to either 120 volts, 208 volts, or 240 volts. The 240-volt connections are at the beginning and end of the primary winding, so there are the maximum number of turns of wire when you use those terminals. The 208-volt connection skips about 10 percent of the turns, and the 120-volt connection is in the middle, using only half the turns.

Both ends of the secondary coil, which together deliver 24 volts

Other end of the primary coil

Ninety percent of the way through the primary coil

Halfway through the primary coil

One end of the primary coil

▽ Transformers come in a wide range of sizes! This tiny thing is used for coupling audio signals. It has the same number of windings on the input and output, so the voltage doesn't change, but it's useful anyway because it completely isolates the two sides from each other for anything other than a changing, alternating current.

I'm not completely sure why the coil is mounted on these plastic springs, but my theory is that this reduces the humming sound that might otherwise be generated by the opposing magnetic fields from the primary and secondary coils (which could vibrate the whole toothbrush if the coil were rigidly mounted to the case).

▽ Wireless chargers for cell phones use the same idea, except the primary and secondary coils are flat, so the phone can be set on top of the charger. They only work when the coils are very close to each other, because the magnetic field around a coil of this shape drops off rapidly with distance.

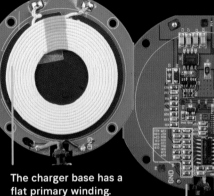

The charger base has a flat primary winding.

Surprisingly complex circuitry (on the back side of the charger coil) is required to detect when the cell phone is placed on the base, and to modulate the power sent to the coil appropriately.

▷ This common form of electric toothbrush charging system is a transformer with one coil in the base, and the second coil in the bottom of the toothbrush. With the case of the toothbrush removed, we can see the secondary coil. The primary coil in the base is difficult to see because it's completely encased in a rubber compound that's been injected all around it for waterproofing. A magnetic iron core sticking up into the bump in the charger guides the alternating magnetic field up from the coil below.

Inside the cell phone, a similar flat coil acts as the secondary.

This is the secondary coil.

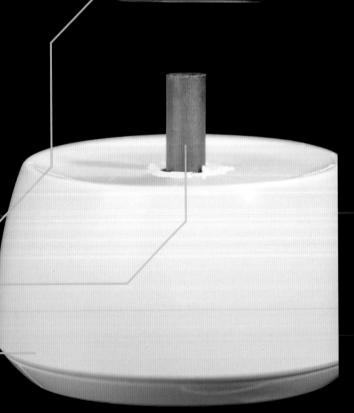

This iron rod extends from the primary coil and draws its magnetic field up.

The primary coil is embedded in this base.

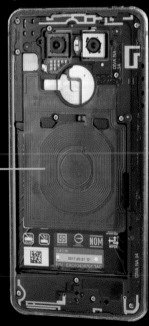

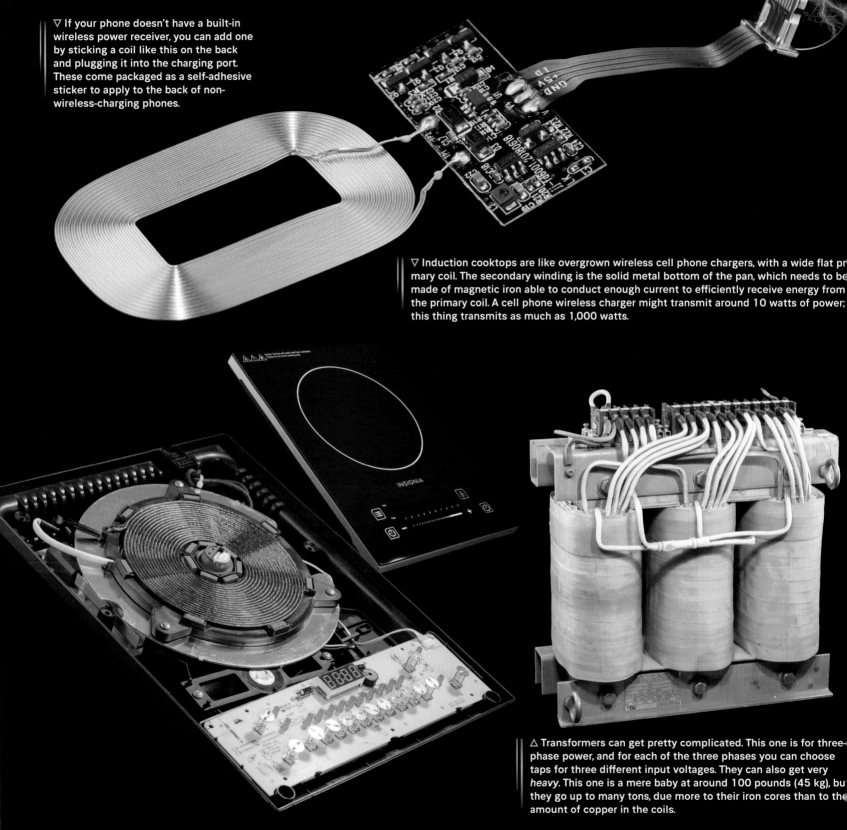

▽ If your phone doesn't have a built-in wireless power receiver, you can add one by sticking a coil like this on the back and plugging it into the charging port. These come packaged as a self-adhesive sticker to apply to the back of non-wireless-charging phones.

▽ Induction cooktops are like overgrown wireless cell phone chargers, with a wide flat primary coil. The secondary winding is the solid metal bottom of the pan, which needs to be made of magnetic iron able to conduct enough current to efficiently receive energy from the primary coil. A cell phone wireless charger might transmit around 10 watts of power; this thing transmits as much as 1,000 watts.

△ Transformers can get pretty complicated. This one is for three-phase power, and for each of the three phases you can choose taps for three different input voltages. They can also get very *heavy*. This one is a mere baby at around 100 pounds (45 kg), but they go up to many tons, due more to their iron cores than to the amount of copper in the coils.

A Motor with No Magnets and No Sliding Connections

EARLIER IN THIS CHAPTER, we left the topic of motors wondering if it was possible to create a motor that has no magnets, only coils of wire, yet doesn't have any sliding electrical connections to bring electricity to the rotating coils. We're going to combine pretty much everything we've learned so far about magnets, electromagnets, generators, and transformers to see how this is possible. Fair warning: this gets a bit weird, but at least we are back to talking about actual motors.

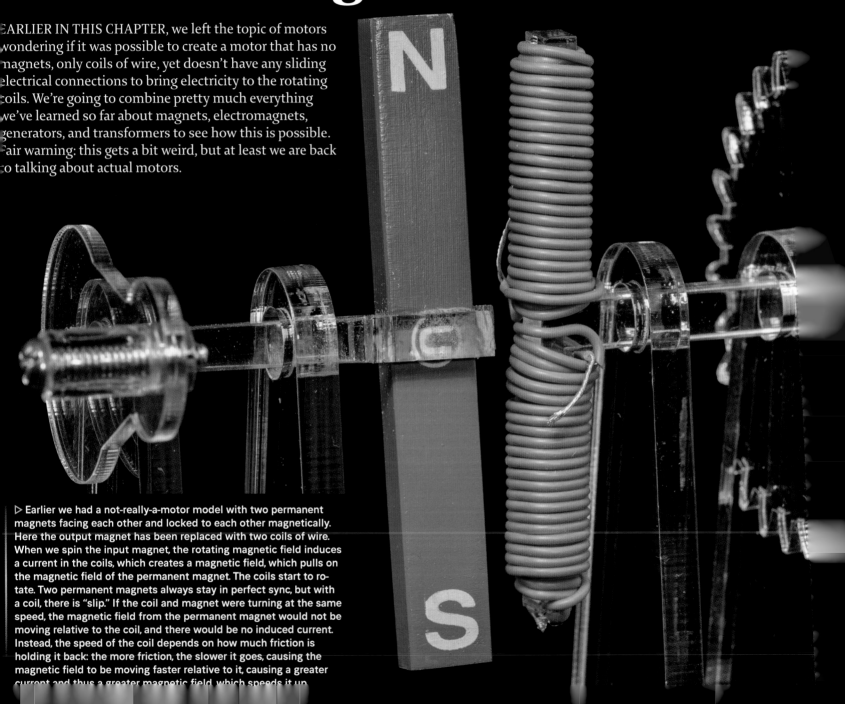

▷ Earlier we had a not-really-a-motor model with two permanent magnets facing each other and locked to each other magnetically. Here the output magnet has been replaced with two coils of wire. When we spin the input magnet, the rotating magnetic field induces a current in the coils, which creates a magnetic field, which pulls on the magnetic field of the permanent magnet. The coils start to rotate. Two permanent magnets always stay in perfect sync, but with a coil, there is "slip." If the coil and magnet were turning at the same speed, the magnetic field from the permanent magnet would not be moving relative to the coil, and there would be no induced current. Instead, the speed of the coil depends on how much friction is holding it back: the more friction, the slower it goes, causing the magnetic field to be moving faster relative to it, causing a greater current and thus a greater magnetic field, which speeds it up.

So far, we know that:

1. Two magnetic fields can push against each other.

2. One or both of those fields can come from currents flowing in coils of wire.

3. A changing magnetic field passing through a coil can induce current in that coil.

4. The changing magnetic field can come from another coil of wire (the primary coil in a transformer).

5. The induced current in the secondary coil also creates a magnetic field, which pushes back against the field from the primary coil.

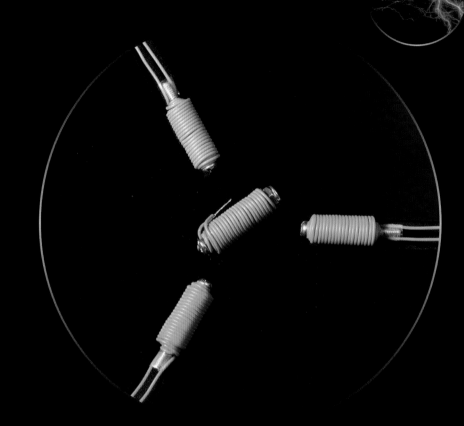

Putting all of this together, we arrive at the fact that if you place two coils next to each other (like in a transformer) and run an alternating current through one of them, they will *physically push against each other*. As long as the second coil has its two ends connected (either shorted directly together or through some kind of light bulb or other load), the alternating current in the first coil will induce a current in the second coil, which will create a magnetic field, which will push against the field from the first coil.

In a transformer these two coils are firmly fastened together, so they can't move. At most you get the sort of humming sound typical of high-power transformers as the coils push and pull against each other.

But here's the clever thing: The second coil only needs to be shorted out, that is, have its two ends connected together. It doesn't need to be connected to anything in the outside world. That means it can act as the rotor of a motor without needing any electrical connection.

Here's the way I think about these motors: the outer, stationary coils are creating a rotating magnetic field, but they are also acting as the primary coil of a transformer. The inner, rotating coil is acting like the secondary coil of a transformer, and because it is short-circuited into a loop, it becomes an electromagnet that is pushed along by the rotating field.

This is very clever. I mean, think about it: This is a device that, when switched off, has no magnetic fields in it at all. There are no permanent magnets, no electrical contacts or switches, no commutator, no transistors, no nothing, just inert coils of wire wound in a particular shape. Yet when you apply an alternating electric current to it, the thing springs to life with multiple interacting magnetic fields that work together to make the shaft spin.

This is so clever, and so effective, that this design, known as an AC induction motor, is the single most common kind of motor in the world. They can be found in large numbers in homes and factories alike. Anywhere there's a motor running on AC power, it's very likely to be some form of induction motor.

△ Here we have the model three-phase motor, but we've replaced the permanent magnet rotor with a coil of wire. The outer three coils, fed with three-phase power, create a rotating magnetic field, and the rotating coil inside follows this field with some "slip" determined by the friction it has to overcome.

A Nice Big Three–Phase Induction Motor

AT FULL LOAD, this 7.5-horsepower three-phase AC induction motor draws 20 amps per phase at 220 volts, which for complicated reasons works out to 7.6 kilowatts of electric power. That's comparable to what an entire home would use with several major appliances on at the same time, even though the motor is only about 18 inches (45 cm) long! Although it is, to be fair, extremely heavy. It took three of us to load it into my van after I got it at auction for the ridiculous price of $20.

The stator coils in this motor are the same as you would find in a three-phase permanent magnet motor. They create a rotating magnetic field that spins at a rate determined by the AC powerline frequency. As in nearly all such motors, there are more than the three poles in the model I showed earlier. Three is the minimum, but multiplying the number of poles brings the ends of the magnets and/or electromagnets closer to each other, increasing the torque and efficiency of the motor. There aren't any more *phases* of power, just more copies of the three phases in multiple sets of coils.

This motor is meant for rough service, so even the cover over the electrical connections is made of thick cast iron.

This ring is for attaching a chain to lift the beast.

Ribs on the case help dissipate heat and make the case stronger.

Because it's an induction motor, as described on the previous page, the rotor contains coils of wire instead of magnets. In practice, the "coils" in nearly all induction motors are not really coils, and they are not made of wire. Instead, highly conductive aluminum is cast into a "squirrel cage" shape within the rotor and completely encased in a stack of thin layers of iron (called laminations) that make up the bulk of the rotor's weight.

The cage works like a coil because current can flow in loops through neighboring rods, connected through the rings on the top and bottom. Functionally the cage is like a set of coils next to each other, each with only a single turn of very thick wire.

In many common squirrel cage designs, the bars and rings of the cage are formed as one single piece of cast aluminum in and around the laminations. This makes for a very strong, integrated rotor, which is important because all the turning force of the motor is created by magnetic interactions happening *inside* these cast aluminum bars, which then push on the laminations, which in turn push on the output shaft.

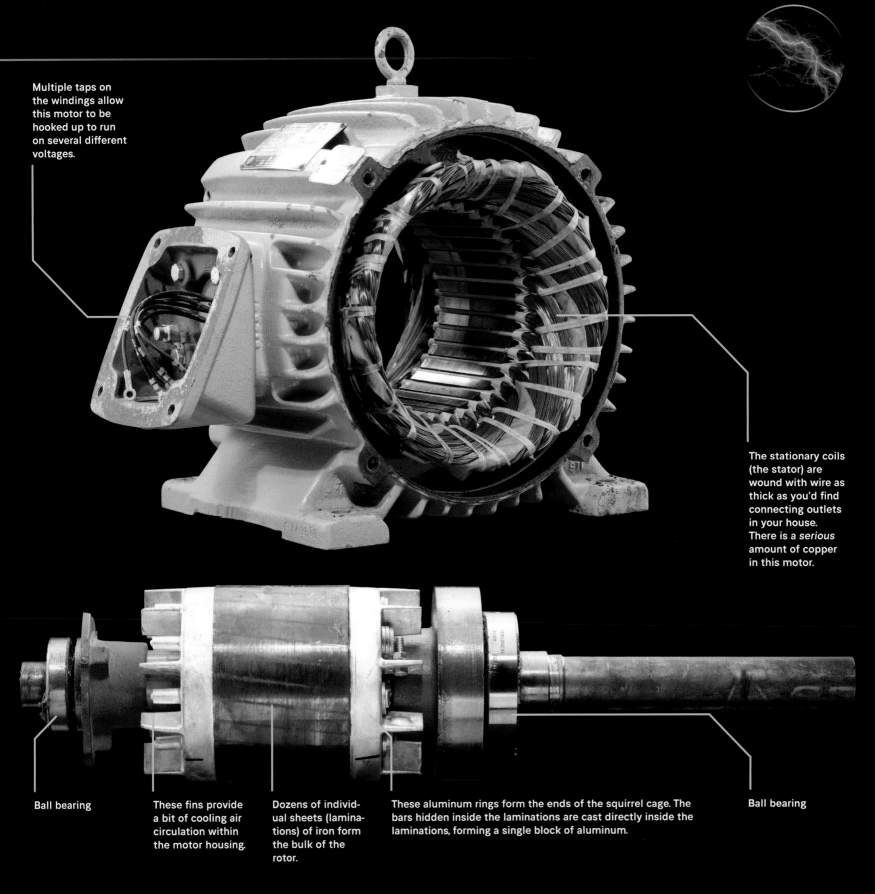

Multiple taps on the windings allow this motor to be hooked up to run on several different voltages.

The stationary coils (the stator) are wound with wire as thick as you'd find connecting outlets in your house. There is a *serious* amount of copper in this motor.

Ball bearing

These fins provide a bit of cooling air circulation within the motor housing.

Dozens of individual sheets (laminations) of iron form the bulk of the rotor.

These aluminum rings form the ends of the squirrel cage. The bars hidden inside the laminations are cast directly inside the laminations, forming a single block of aluminum.

Ball bearing

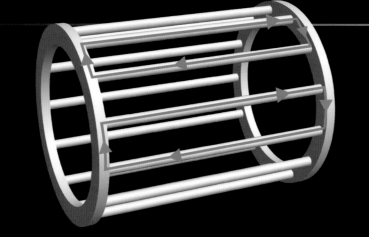

▷ Inside every "squirrel cage" induction motor, usually impossible to see, is an arrangement of copper or aluminum similar to this diagram. As you can see, it would not actually work well for keeping a squirrel, because these animals are smart enough to figure out that both ends of the cage are completely open. But it will hold current, allowing it to flow in loops as shown.

▷ This squirrel cage is unusual because it's fabricated from blocks and bars of copper instead of being cast in place, which is why it's possible to take a photograph of it. Notice just how *thick* these bars are, easily heavy enough to carry hundreds of amperes of current. This rotor was designed as part of the European ReFreeDrive project to create an induction motor with the necessary power-to-weight ratio for use in an electric car. This would have the advantage of replacing the neodymium-iron-boron permanent magnets used in most current electric car motors. (Neodymium is one of the class of elements known as "rare earths"; these elements are not actually *very* rare, but replacing all the cars in the world with ones with permanent magnet motors would create a significant strain on the world's supply of neodymium. Induction motors with copper rotors could be cheaper and less reliant on problematic metals.)

▷ Squirrel cages can also be cast in place using copper instead of aluminum. Copper is a superior material because it conducts electricity better than aluminum, but its melting point is much higher, making it harder to work with. This cast copper squirrel cage was made by the Breuckmann company in Germany for the ReFreeDrive project mentioned earlier. The iron laminations have been machined away to reveal the shape of the copper bars. Is there a material that would be even better than copper? Yes! Silver has an even higher conductivity, but unfortunately it is a bit too expensive. Beyond that, the only way to get a better electrical conductor is with superconducting materials. Motors with superconducting rotor coils have been built, but for now they are just prototypes or test units, and they are *very* big—1,000 horsepower or more.

▽ This motor, which powers a drying fan on the side of a grain elevator (a large bin used to store corn or soybeans), is also a squirrel cage induction motor, and looks a lot like the one on the previous page, but a step up in size. It's rated at 50 horsepower and weighs 620 pounds (280 kg), according to its data tag. These motors are incredibly robust and reliable even under harsh environmental conditions, because they have essentially no active components, and no moving parts other than the rotor spinning in the middle. There's just very little that can go wrong with a motor like this.

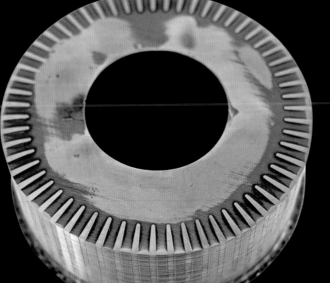

Bigger Induction Motors

VERY LARGE INDUCTION motors, like the size of a car or bus, usually have actual coils of wire in their rotors instead of squirrel cages. And they bring out the ends of those coils to a set of slip rings, which let you connect to them outside the motor. If these slip ring connections are short-circuited to each other, the rotor works just like a squirrel cage (where the "coils" are internally short-circuited). That is exactly how these motors are typically run—with the slip rings shorted out. But it's not how they are *started*.

When you switch on the power to an induction motor, the rotor is not moving (obviously, because you just turned it on). The magnetic field around it, however, is going full speed, at 1,800 rpm for a typical 60-hertz motor. That means the field lines are cutting through the rotor coils at a tremendously high speed, creating a *huge* current in the rotor. This in turn creates a huge magnetic field in the rotor, which gives the motor a tremendous amount of starting torque (the ability to get going from a cold start).

That's great in some applications, such as electric cars or train locomotives, but for a 5,000-horsepower motor in a factory, it's likely to be so much current that it would overload the factory's electri-cal system, and possibly burn out the motor. The solution is to insert a certain amount of electrical resistance in series with the rotor coils, by way of the slip ring connections. This limits the amount of current to something manageable. As the motor comes up to speed, the amount of resistance can be progressively lowered until it's eventually eliminated, meaning the coils are shorted out. When the motor is running at its full intended speed, the rotor is turning almost as fast as the magnetic field spinning around it, and the current flowing through the rotor coils is only just enough to keep it going.

▽ Slip ring induc-tion motors tend to be very large. This rotor alone weighs as much as several cars.

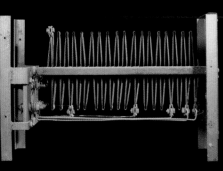

△ The sort of resistor used to limit the current in a large slip ring induction motor is itself large, because when current runs through a resistor, heat is generated, which needs to be dissipated somehow. This old one, from a theater arc light's motor-generator, looks more like an electric heater than a resistor—which is no surprise. The only difference between a big electric resistor and an electric heater is your attitude toward the heat it gives off. If you are interested in its ability to limit the flow of current, then it's a resistor (and the heat is an undesirable side effect). If you want the heat, then it's a heater (and the fact that it's a big resistor is just the means to the end).

△ Motors and generators can get very big. The largest nearly all have wound rotors with slip ring connections because otherwise starting them up would be nearly impossible.

△ The heating element inside this cheap space heater is nothing more than a big electrical resistor, which gets hot when current is run through it.

▷ Peering into the end of the teaching motor, we can see the stator coils, rotor, and slip rings.

Graphite brushes that make a sliding connection to the slip rings

Wound rotor coils, which spin when the motor is running

Copper slip rings

Stationary outer coils (stator)

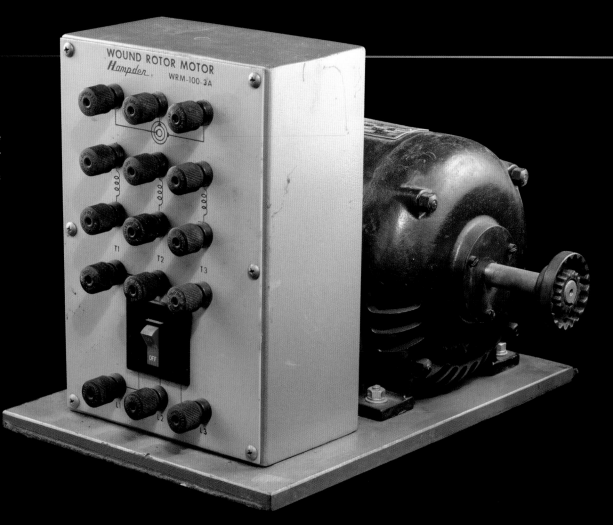

▷ This three-phase slip ring induction motor is *much* smaller than most motors of this type—just ⅓ horsepower. That's because it's a teaching motor designed to let students experiment with this type of motor in a reasonably sized classroom. The ends of all its coils, both stationary and rotating, are brought out to terminals on the front panel, where the student can experiment with connecting them in different ways.

▷ On the front panel the student can use lengths of wire, resistors, or other electrical components to experiment with making the motor run. Hopefully they are careful and read the textbook first, because this motor is designed to run on 208 volts, and things could get very exciting if it's wired wrong! It can also run on much lower voltage, with correspondingly reduced power—which is probably a good idea for beginning students.

These three terminals connect to the three slip rings, each of which is wired to one end of one of the three rotor coils. (The other three ends of all the coils are wired together in the center. This is called a Y configuration: the three slip rings are connected to the ends of the three arms of the Y.)

These six terminals let you connect the three stationary coils any way you like—including ways that will make the motor run forward, backward, or not at all.

This switch disconnects all three phases, making the rest of the panel safe to work on.

Incoming three-phase power is connected to these three terminals.

When the switch is on, three-phase power is present at these terminals.

As we've learned, when induction motors are starting up, they draw vastly more current than when they are running at full speed. The good thing about this is that they can supply a large amount of torque right from a standstill, which is very useful for getting heavy objects into motion, like a locomotive pulling a long train. The downside is that whatever is supplying power to the motor has to be able to supply not only the normal running current, but also the peak starting current.

If you have a whole lot of electric motors all connected to a single generator, you can get away with a much smaller generator if you're certain you won't ever have to start up all the motors at the same time. For this reason, it is customary to stage the startup of very large electric motors, with enough delay between the stages to protect the generator.

△ Rules for starting induction motors apparently apply even when the motors are completely imaginary, as in this computer-generated scene from the movie *2012*. Here we are seeing one of the giant imaginary ships carrying the imaginary remnants of humanity after the world has been flooded by some kind of issue with the sun. It's the end of the world/movie, all is well, and they are opening the hatches for the first time. From the fact that the opening is staged, as seen in the picture, we can conclude two things. First, that the imaginary onboard generators are able to supply enough power to operate all the hatch doors at the same time, but *not* enough power to start them all simultaneously. And second, that the art director knew it would look way cooler to have them open in a wave like the wings of a bird in flight.

When Three Phases Are Too Much

THREE-PHASE MOTORS are beautiful, but they require three-phase power, which is awkward and expensive to deliver: It requires three rather than two wires (or as many as five, if you want three phases plus a neutral and ground). That's worth it for large motors, but for motors of less than a few horsepower, there are other designs that work very well on the single-phase AC power you get from an ordinary wall plug.

I started my explanation of AC motors with three-phase motors because, despite the extra complexity of three-phase power, they are the easiest to understand. Once you understand the idea of a rotating magnetic field, the rest starts to make sense. But if all you have is single-phase power, it's more difficult to get a rotating field out of it.

To see the problem with single-phase power, consider what happens if we take a three-phase motor model and remove two of the coils, leaving just a single coil being driven by a single AC sine wave voltage. The magnetic field created by this coil will flip back and forth 60 times per second, alternately pushing and pulling on the rotating magnet. It *is* possible for this to run as a motor, but there are problems.

First, it will only work if the magnet is already spinning exactly in sync with the AC frequency . . . and there is no way to get it up to that speed initially. If you just switch on the current, the magnet will sit there and jitter back and forth without turning in any direction.

Second, if you do get it started, it will run in either direction equally well. This is a bad thing: in almost all situations you want a motor that will always run in the same direction, not start randomly in different directions, or have to be manually started in the direction you want it to go.

Third, when the motor is running, the force turning the shaft will pulse with every cycle of the AC current (120 times per second if you have 60 Hz AC power). This makes the motor noisy and adds wear and stress to whatever the motor is driving.

A few motors with this design do exist, but only in limited, low-power applications. They typically have some mechanical ratchet to make sure they can only rotate in one direction. This ratchet can also turn the initial jittering into rotation in the desired direction, as long as there is little or no load on the motor when it's started. To make a more practical motor, we need some way to get the magnetic field to rotate, not just flip back and forth.

Shifting Phases

IF WE CAN'T HAVE a perfect rotating magnetic field from two-phase or three-phase power, then maybe we can create a sloppy, sort-of-rotating field from single-phase power. How would we go about doing that?

An Etch A Sketch makes for a good analogy. This classic toy lets you draw pictures on a screen using two knobs. The left knob moves the line side-to-side, and the right knob moves it up and down.

Imagine that you have an Etch A Sketch, but you're only allowed to turn the left knob. No matter what you do, the pointer is always going to go left and right in a straight line. This is analogous to the magnetic field created by a single coil of wire: no matter what you do, the magnetic field it creates will always point along the axis of the coil.

Left knob

Right knob

△ I've made two graphs that show the positions of the left and right knobs as a function of time. In this example, the left knob is turned steadily clockwise (shown by its curve rising). The right knob stays unchanged (shown by its curve staying flat).

△ Now imagine you're allowed to turn both knobs, but there is a belt connecting them so they both have to turn the same amount at the same time. Now the pointer will move both up and down and left to right, but it will still be a straight line, just at an angle. The magnetic analogy is that if you use two coils of wire pointing in different directions near each other, both wired to the same AC current, you can still only create a magnetic field that varies along a

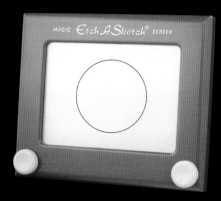

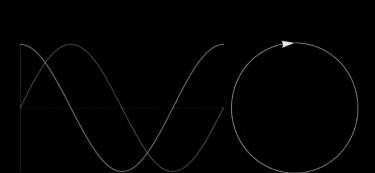

△ Here are the two circle-creating curves on one axis. Both of them are the same shape, but one is shifted in time by a quarter of a cycle, which represents 90 degrees around the circle they are tracing out. These are generically called sine waves, and because they are 90 degrees shifted, or "90 degrees out of phase," it's customary to call the one that rises first a sine wave and the other one a cosine wave. You will notice that this diagram is identical to the two-phase diagram and the diagram that shows the relative motions of the piston and valves in a simple steam engine. Sine and cosine waves are very closely related to circles, which is of course why they're ever-present when creating a rotating magnetic field.

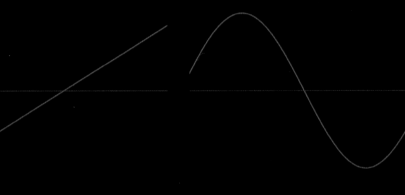

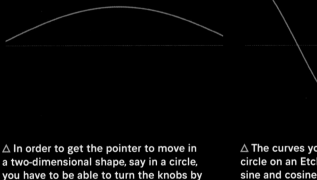

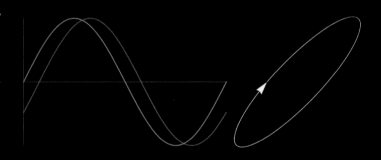

△ In order to get the pointer to move in a two-dimensional shape, say in a circle, you have to be able to turn the knobs by *different amounts at different times*. For example, if you move the pointer steadily upward with the right knob, while using the left knob to move it first to the left and then to the right, the pointer will move through an arc as it's rising. It's not a circle yet, but at least it's not a straight line.

△ The curves you need to create a perfect circle on an Etch A Sketch are, no surprise, sine and cosine waves. These curves, and only these curves, combine to move the pointer smoothly in a circle at a constant speed. The only other way to get a perfect circle on an Etch A Sketch is to fake it with Photoshop. I'll let you guess which way I did it.

△ If, instead of sine and cosine waves 90 degrees out of phase, two sine waves are shifted by a smaller amount, the result is an ellipse instead of a circle. For purposes of running an induction motor, this isn't as good as a true circle, but it's way better than the straight line created by only one phase. In particular, even a very small amount of rotational component to the field is enough to get the motor turning in the right direction, and once it's up to speed, it will run reasonably well even with just a single phase. So, a single-phase AC induction motor doesn't necessarily have to run with a perfect 90-degree shifted phase. It just needs something to shift phase a little bit, at least until the motor is running at full speed.

A common way to add a second, slightly shifted phase is using a capacitor. This is a device that can temporarily store electric charge, acting a bit like a rechargeable battery that can be charged and discharged *very* rapidly. When you put a capacitor in series with a coil of wire, and then supply it with alternating current, the capacitor charges up while the supplied voltage is high, then discharges when the supplied voltage goes back down again. The net result is to delay the point of maximum and minimum current flowing through the coil, in effect shifting the phase of the current in exactly the way we want. We have now constructed what is called a capacitor-run single-phase AC induction motor.

This is not a hump, it's a square box in which you make the wiring connections. It's common to find several more wires than you would need; these are different taps on the motor windings, which allow you to run the motor on different AC voltages (or burn it up, if you connect them wrong).

▷ Inside the single-hump motor above, we find a "squirrel cage" induction rotor, and a very nice armature (the stationary windings that surround the rotor).

These stator coils don't move and are connected to the power supply through an appropriate set of wires depending on the voltage you're running it on.

▷ Capacitors work by having two plates of metal very, very close to each other. When you "charge up" a capacitor, electric charge builds up on the two plates, positive on one side and negative on the other. Because positive charges are attracted to negative charges, the two sets of charges hold each other together, and can be stored there for a long time. Or, they can be drawn back out again a fraction of a second later.

This hump contains a capacitor.

A motor with a rounded hump on the side is almost guaranteed to be a single-phase AC motor. The hump is the capacitor used to create the necessary second phase. It can be either a "run capacitor," which means the capacitor is always engaged in the circuit and the motor runs as described above. Or it can be a "start capacitor," which means the capacitor is connected only while the motor is starting up. Once it's up to speed, a centrifugal switch activates and disconnects the capacitor, after which the motor runs on just a single phase. (Again, it's possible to *run* a motor on only one phase, you just can't start it—hence the temporary need for a capacitor.)

This is a switch that disconnects the capacitor from the circuit once the motor reaches a defined speed. It works just like the governor of a steam engine, except instead of throttling back steam as the engine speeds up, the spinning weights snap an electrical switch once they are going fast enough to create sufficient outward pressure. (The switch itself doesn't rotate, so there's no need for a sliding electrical contact.)

Much like camels, single-phase AC motors come in single-hump and dual-hump varieties. As we just saw, single-hump motors have a single capacitor that could be either a run or a start capacitor. Dual-hump motors use one capacitor to get started, and a second one while running at their operating speed. The two capacitors are different in design: The start capacitor is optimized for high current, to deliver starting torque, but would overheat if left connected all the time. The run capacitor has a lower capacitance, but is able to operate continuously.

Capacitor

Capacitor

Wiring box

This teaching motor has the connections to its two coils brought out to the front so the student can experiment with different ways of connecting it to see if they can get it to run.

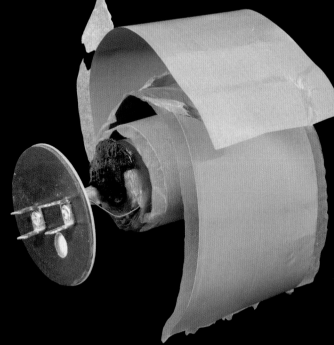

△ To make a capacitor able to store a lot of charge, you need a large surface area of metal, and you need the two plates to be very close to each other. The normal solution is to wrap two long strips of metal, separated by a thin insulator, into a coil.

These two terminals connect to the out-of-phase starting coil.

These terminals connect to the main stationary windings.

When the switch is on, these terminals are powered up.

This indicates that the coils will be automatically disconnected by a centrifugal switch when the motor has finished starting up.

These terminals need to be connected to 120-volt single-phase power.

This switch is also a circuit breaker, in case the student makes a mistake.

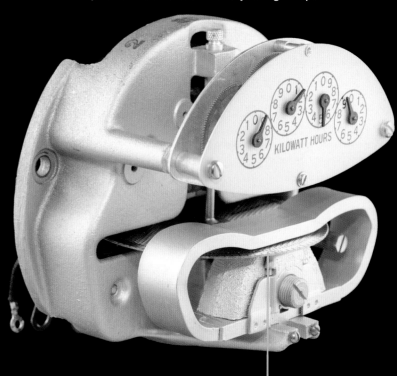

◁ Electric power meters measure how much power a house is using so the power company can charge you an appropriate amount for it. (Or, in this case, the meter measures how much power one of two different light bulbs is using, so you can compare how efficient they are.) Surprisingly, these meters are essentially AC induction motors of a very strange shape.

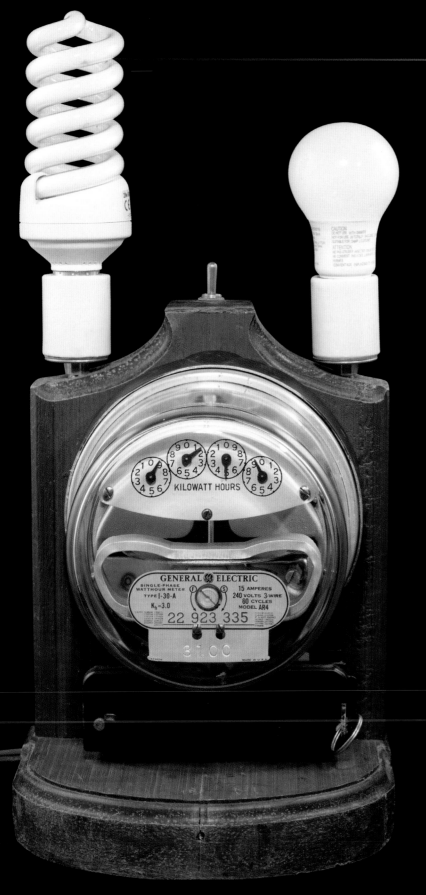

△ From the front you see a flat aluminum disk, which spins faster the more power you are using. Dials with delicate brass gears add up the total number of times the wheel has turned, which tells the power company how much to charge you.

This aluminum disk is actually the rotor of an induction motor.

▷ From the back we can see the stator coils of the induction motor that is this meter. There are two sets of coils that work together to make the meter run at a speed proportional to the total power running through it. The coil above is connected directly across the incoming power lines, and has many turns of fine wire. The coils below have a small number of turns of very thick wire (more like flat copper bars), because they are wired in series with the house and have to be able to carry the full amount of current, as much as 200 amps, flowing into the house. Together these coils generate an alternating magnetic field through the aluminum rotor disk that passes between them. This creates a current in the disk, which creates a magnetic field, which pushes against the coil's field, making the disk rotate. The coils on the bottom create a force proportional to the current flowing through the house, and the coil on the top creates a force proportional to the supply voltage. Power equals current times voltage, so between the two coils, the force in the disk is proportional to power.

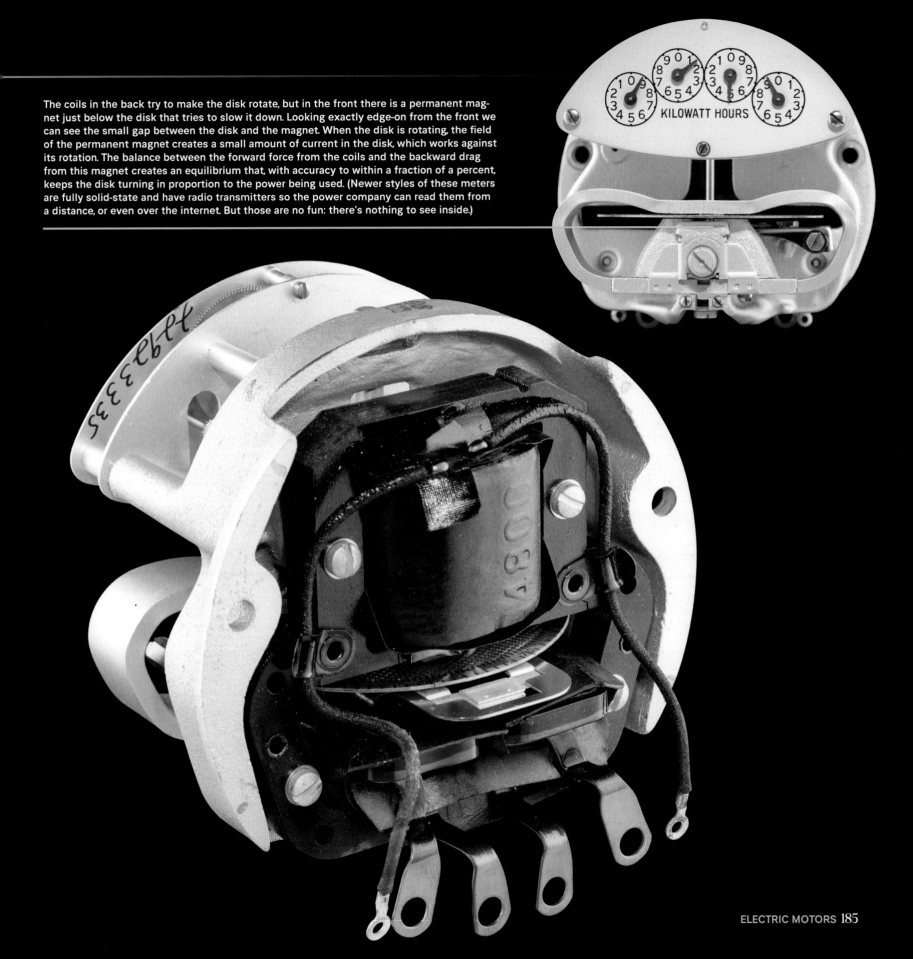

The coils in the back try to make the disk rotate, but in the front there is a permanent magnet just below the disk that tries to slow it down. Looking exactly edge-on from the front we can see the small gap between the disk and the magnet. When the disk is rotating, the field of the permanent magnet creates a small amount of current in the disk, which works against its rotation. The balance between the forward force from the coils and the backward drag from this magnet creates an equilibrium that, with accuracy to within a fraction of a percent, keeps the disk turning in proportion to the power being used. (Newer styles of these meters are fully solid-state and have radio transmitters so the power company can read them from a distance, or even over the internet. But those are no fun: there's nothing to see inside.)

KILOWATT HOURS

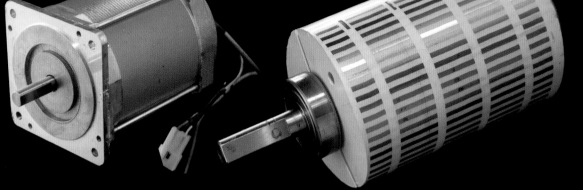

△ In the armature there are twelve separate coils, wired in two groups. If you send current to one pair of wires coming out of the motor, all the even-numbered coils energize. If you send current to the other pair, the odd-numbered coils energize. Each coil concentrates its magnetic force on a set of four teeth, giving a total of forty-eight teeth. Notice that this is two fewer than the fifty teeth in the rotor. The spacing of the teeth within each group of six exactly matches the spacing of the teeth in the rotor, and each group of six is offset by one-sixth of a tooth-space, so over the twelve groups the missing two teeth are accounted for. This strange spacing is what allows the motor to incrementally step by one tooth-width per cycle of the input power.

△ This Slo-Syn brand motor has fifty separate magnetic poles, so each cycle of the AC voltage turns the permanent magnet rotor only $1/50$ of a revolution. Operated on 60-cycles-per-second AC power, that makes this a 72-revolutions-per-minute motor (which is a little over one revolution per second). A capacitor is required to provide a second phase if you want it to run continuously. Alternately, you can control the two phases independently with DC current, which in effect turns this motor into a stepper motor.

△ From the side it's hard to see anything special about the arrangement of magnets on the surface of the rotor, but if you sight down it lengthwise, you can see that they are arranged in sets that alternate like a stretched-out checkerboard. These are alternately north and south poles.

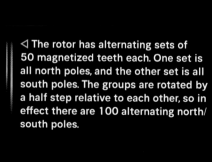

◁ The rotor has alternating sets of 50 magnetized teeth each. One set is all north poles, and the other set is all south poles. The groups are rotated by a half step relative to each other, so in effect there are 100 alternating north/south poles.

△ I have a particular fondness for Slo-Syn motors because back in high school, around 1980, my friend Donald Barnhart and I built this robot, which is operated entirely by Slo-Syn motors exactly like the one above. It was pretty intense. I built the electronics and the handheld control panel, and he welded the frame. We didn't have access to a microcomputer, so it was entirely hand-guided, but it was brutally strong.

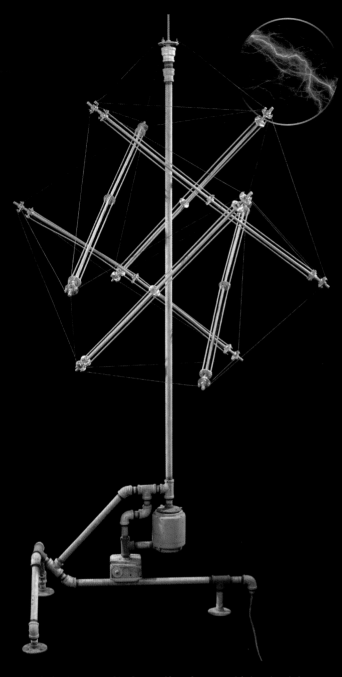

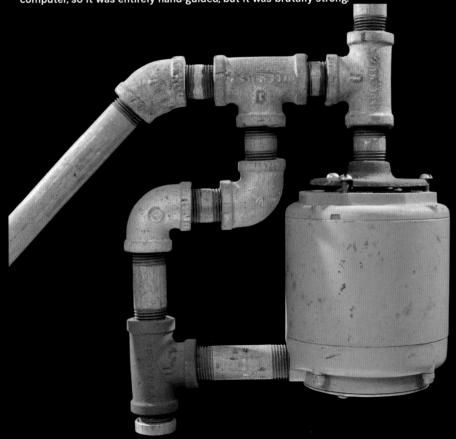

△ This is the other Slo-Syn artifact from my high school days. It's a tensegrity figure, that is, a structure that uses both tension and compression elements to create a shape. As you can see, the piece dates from my plumbing phase as an artist. I call it *Shiva, the Destroyer of Space* because it is impossible to find any place to put it where it's not in the way. When I dug it out to take pictures for this book it had not been plugged in for probably going on thirty years, but such is the nature of electric motors that I had zero doubt it would start immediately and run without problems, which it did.

◁ A Slo-Syn motor in the base turns a long shaft that goes up through the support pipe. At the top a ball bearing separates the rotating shaft from the sculpture. After a few minutes, friction in the bearing brings the whole thing into 72-rpm rotation, which is actually pretty fast for something this big.

▽ This is an example where the fact that the motor can turn only at a rate exactly in sync with the power line is an advantage. In fact, it's the whole point: This is called a synchronous timing motor. Because the AC power you get from the wall runs at a very precise frequency, so will this motor. You could use one to run a clock and it would be as accurate as any clock you can buy. In fact, exactly this brand of motor is used in some larger or custom-made clocks.

▽ In our studio we shoot a lot of "rotations" of objects. That means we set them on this turntable, which uses one of these timing motors to rotate the object around a complete circle in *exactly* twelve minutes. A camera pointed at the object uses a time-lapse controller to take a picture *exactly* once per second. At the end of twelve minutes, the camera and turntable are still perfectly in sync, creating a seamless set of 720 photos at $1/2$-degree intervals.

△ The rotor of a timing motor is a cup-shaped piece of magnetized iron that is driven from the inside by coils.

▷ All motors of this style have rotors inside that turn at the same rate, but you can buy them with output shaft rotation rates of anywhere from one revolution per second to one revolution per *week*. A particular combination of gears translates the rotor speed to the output speed.

△ This tiny permanent-magnet synchronous motor is from a battery-operated wall clock. Because it's run on a battery instead of AC wall current, it needs a quartz crystal oscillator to create AC current of a very precise frequency.

Confused Motors

WE'VE SEEN HOW internal combustion engines use pistons and connecting rods to push on a crankshaft, making it turn. And we've seen how electric motors use a smoother, continuous circular pushing/pulling arrangement without any connecting rods or crankshaft. But could you switch this around and make an internal combustion engine with no crankshaft, or an electric motor with one?

Yes, you can! But people generally don't, because neither idea works as well as the conventional way of doing it.

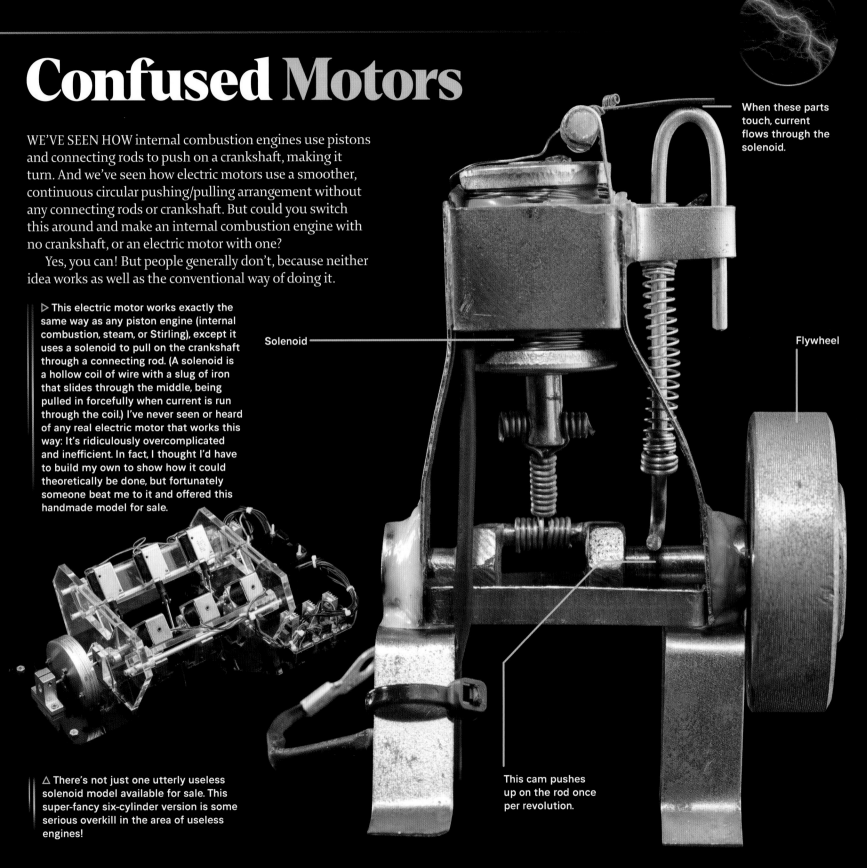

▷ This electric motor works exactly the same way as any piston engine (internal combustion, steam, or Stirling), except it uses a solenoid to pull on the crankshaft through a connecting rod. (A solenoid is a hollow coil of wire with a slug of iron that slides through the middle, being pulled in forcefully when current is run through the coil.) I've never seen or heard of any real electric motor that works this way: It's ridiculously overcomplicated and inefficient. In fact, I thought I'd have to build my own to show how it could theoretically be done, but fortunately someone beat me to it and offered this handmade model for sale.

When these parts touch, current flows through the solenoid.

Solenoid

Flywheel

This cam pushes up on the rod once per revolution.

△ There's not just one utterly useless solenoid model available for sale. This super-fancy six-cylinder version is some serious overkill in the area of useless engines!

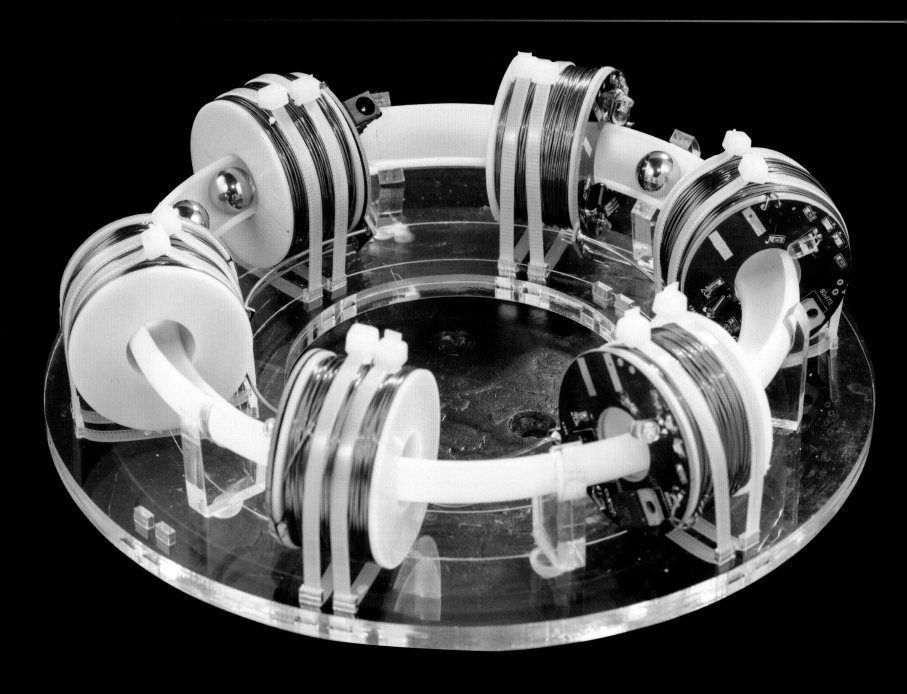

△ This has got to be the worst, and coolest, electric motor of all. It's a ball bearing synchrotron, much like the Tevatron particle accelerator at Fermilab or the Large Hadron Collider in Geneva, except smaller and about a hundred million times cheaper. It has a series of magnets, just like a particle accelerator, and a circular track, also just like a particle accelerator. But instead of something exotic like antiprotons circulating in the track, it's steel ball bearings. Sensors detect when each ball is approaching a magnet, which is then switched on to give the ball a kick and keep it going around the track.

▷ The Wankel engine, made famous by Mazda, is the closest thing to an internal combustion engine that works like an electric motor. It has no crank arms or connecting rods; instead, the force created by exploding fuel is applied directly to the output shaft through a triangular cam. This design is certainly more practical than the electric piston engine above. Millions were sold, and they have some real advantages (including high power-to-weight ratio, good for the sports cars Mazda put them in). Unfortunately they have bigger disadvantages, including low fuel efficiency and sky-high pollution emissions.

A Design That People Keep Thinking Should Be Good for Something

SOME ELECTRIC MOTORS, while still based on magnetic fields, take the idea off in strange directions. For example, the very first electric motor, designed by Michael Faraday in 1821, was as simple as it was insane (at least by today's standards).

It's called a "homopolar" motor, meaning "same polarity," because the polarity (direction) of the current flowing through it stays the same all the time. It's the only kind of motor that truly runs on DC and not some form of AC created either inside the motor with a commutator, or outside the motor with some kind of circuitry.

I went to some trouble to create a version of Faraday's original design, which relies on a large pool of liquid mercury. Mounted in the center of the pool is a permanent magnet. Hanging from above is a copper rod, free to swing as it pleases. Connected between the top of the rod and the pool of mercury is a battery, which sends a strong electric current through the rod.

With the battery connected, the rod rotates slowly around the magnet, propelled by the circular magnetic field created by the current running through it. In theory you could extract some mechanical work

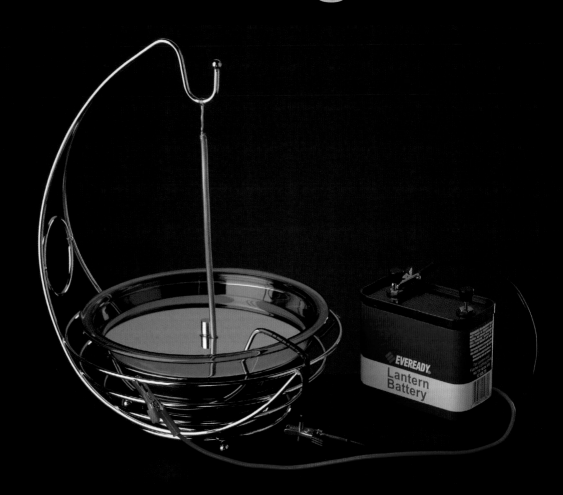

from this rotation, but only a tiny amount compared to how much power is coming from the battery. Also, mercury is terribly poisonous and really shouldn't be kept in open pools like this!

△ This design is essentially identical to Michael Faraday's original, except he didn't use a banana stand to hold up the rod. I feel the stand is appropriate, considering how completely bananas this motor is.

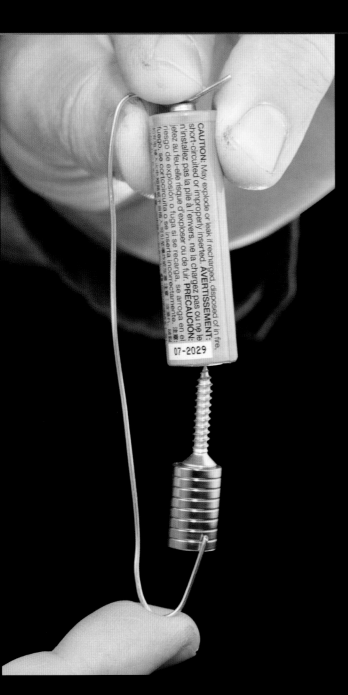

△ Here is a nontoxic, handheld version of the same idea. The stack of disk magnets lets the steel screw hang magnetically from the bottom of the battery, while the length of bare wire completes a circuit through the magnets. It's a bit tricky to hold everything just right, but when you do, the magnets spin quite fast. Despite their problems, people did try making real homopolar motors, but gave up something like a hundred years ago. In recent years there has been some renewed interest in maybe making them using superconducting coils, but it's not looking like the technology of the future.

△ As with most electric motors, homopolar motors also work in reverse as generators. Surprisingly, these have actually found a few applications in modern times, because they have an unusual ability to produce huge amounts of current (at low voltage) for short periods of time. This makes them interesting for doing things like powering a rail gun, which requires a pulse of incredibly high current (a couple million amps) to throw a projectile at speeds faster than possible with any gun based on chemical explosives. To create such high-intensity pulses, homopolar generators store up a lot of energy in a very heavy spinning iron disk (this one was taken out of a research generator and made into art). Once the disk, spun up by an ordinary electric motor, is going very fast, the circuit is closed and current is drawn off from the disk through sliding contacts. The current can be very high because the internal resistance of the thick iron disk is very low.

Raw Power at the County Fair

COUNTY FAIRS HOLD a special place in the rose-colored landscape of rural and rural-adjacent America, with their blue-ribbon pies, extremely loud tractor pulls, food you seriously regret eating, and carnival rides that double your regret.

What I find most fascinating about county fairs is the raw, unfiltered electricity that runs through them. We're not talking about dainty sips of current flowing in neatly organized wires or pulsing through delicate silicon chips. Here there are great whopping buckets of electrons being pumped recklessly through sausage-thick cables laid in the grass, spreading out to rides up and down the midway. Big brass contacts spark

and chatter as motors are switched on and off. The entire ceiling of the bumper car ride pavilion is electrified with potentially lethal voltage, which is picked up and brought down to ground level by "stingers" on the back of each car.

For reasons I don't entirely understand, carnival rides are typically powered by large truck-mounted diesel generators stationed very near or even within the midway. This is done even at fairgrounds—like the one in my town—that could easily be supplied plenty of three-phase power from the local utility. These generators add a droning background noise matched only by the demolition derby and the country music act on Saturday night.

Running the various rides are an assortment of single- and three-phase AC induction motors of the kind we just learned about. Ranging up to dozens of horsepower, these beasts run quietly, cleanly, and reliably. If it weren't for the generators, the visitors, and the country music act, the whole place could be nearly silent. You can actually experience this silence in places like the indoor carnival at the Mall of America in Minneapolis, where premium-quality versions of the same rides run on smooth rubber wheels, and all you hear is happy music and kids screaming (the parents are screaming on the inside, so you don't hear them).

◁ This scene hasn't changed in fifty years, because these are the same rides as fifty years ago. Not the same kind, the same actual rides—and I can prove it.

▷ Foods like these are typical at Midwestern county fairs.

◁ When I was a little kid some fifty years ago, this ride was called the Starship 2000, because that was as far in the future as anyone could imagine—a time when we would all have flying cars. (I rode it once and it gave me a splitting headache.) When it came time, some thirty years later, for my kids to go to the county fair for the first time, the *exact same ride* was still there, now named the Starship 3000. How did I know it was the same ride? Because for several years you could see the outline of a 2 in faded paint underneath the 3 in 3000. It's still there in 2021; having learned their lesson the first time, they clearly decided to leave an extra margin of safety in the name.

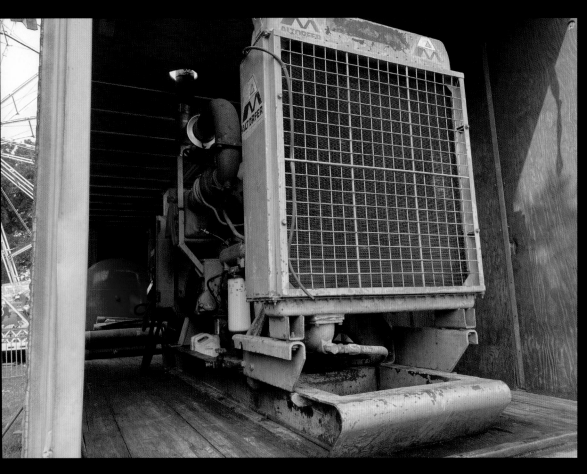

△ Several large diesel generators mounted in the backs of truck trailers run nonstop while the fair is in session.

▷ Like snakes in the grass, thick electrical cables offer a powerful bite if provoked. They run from the generator trucks out to each of the rides and stalls. People just walk over them. It's kind of wild.

△ This kid-sized Ferris wheel has a relatively small AC induction motor, just a few horsepower, coupled to the Ferris wheel through a simple enclosed gearbox. (Ferris wheels are named after George Washington Gale Ferris Jr., who invented them. If they had been named for the fact that they are made of iron, they would be called ferrous wheels.)

△ This larger Ferris wheel has, naturally, a larger motor. Instead of an enclosed gearbox, there are belts and pulleys to slow the typical speed of 1,720 rpm down to about 1 rpm of the wheel. The belts and pulleys are hidden behind a safety cage because they tend to catch fingers or clothes and thrash them violently.

▷ This is my favorite ride. Not because I'd actually want to go on it—I hate these things— but because it has the best and most visible mechanism. What a beautiful motor and gearbox!

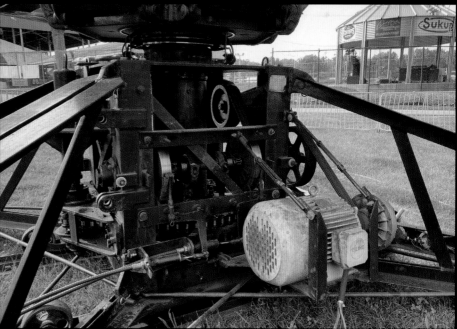

▽ Bumper cars are where electricity is most exposed. The metal floor is grounded, while the metal ceiling is live with at least several dozen volts. (Different models and eras of bumper cars use anywhere from 24 to 110 volts, some AC and most DC.) Higher voltages allow for more powerful cars with less current to wear out the contacts, at the cost of a greater chance of killing the guests.

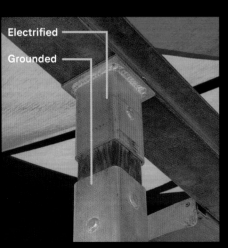

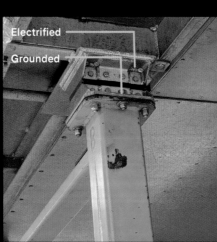

Electrified
Grounded

Electrified
Grounded

△ Each car has an electric motor, brushes that make contact with the metal floor, and a pole, called a stinger, that reaches up to make contact with the ceiling. The pole is grounded on the outside with an insulated wire running inside, so it won't shock you if you touch it.

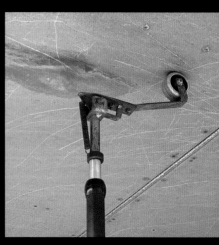

△ Wood and rubber insulators electrically isolate the ceiling of the bumper car pavilion from the rest of the structure. I'm a little surprised at the use of wood, since it is not a good insulator when wet, but I guess it works well enough to meet county fair electricity containment standards.

△ A metal wheel rolls on the ceiling, completing the circuit to run the car. There are sparks every time it crosses a joint in the ceiling.

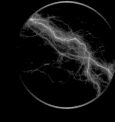

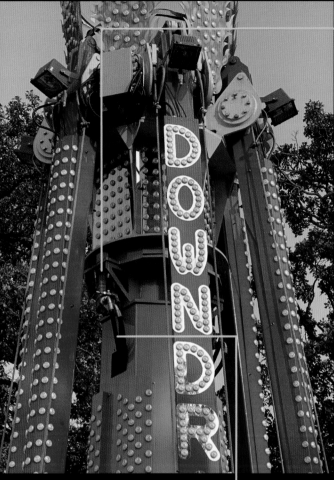

Notice how small this motor is compared to the electric motors we've seen on the other rides—yet it's driving the main rotation of quite a large ride.

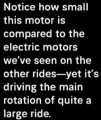

Hydraulic pump

Electric motor

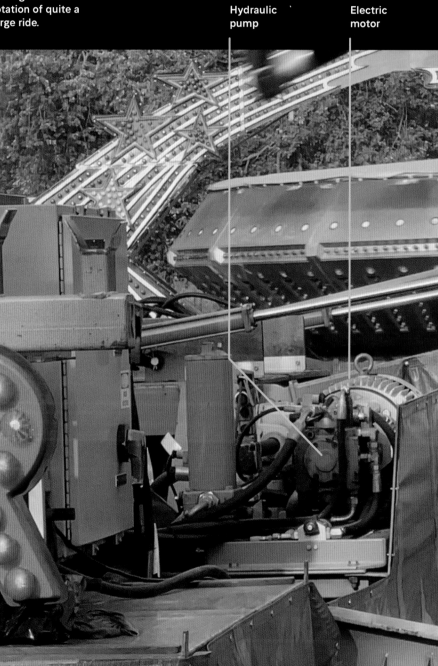

△ Some rides are driven by hydraulic motors rather than electric. Hydraulic motors are higher-maintenance than electric, but they allow a tremendous amount of force to be applied in a specific location with a very small device, and they are just as quiet.

High-pressure hydraulic hoses

▷ Rides that use hydraulic motors are still electrically powered. They just have an electric motor hidden in the base somewhere driving a hydraulic pump, which delivers high-pressure hydraulic fluid to the motors.

△ This ride, named Downdraft, bounces its passengers repeatedly up and down in a matter of seconds, while spinning them around. Leaving aside the question of why anyone would want to, there's a problem making a ride like this work: No motor of a remotely practical size could deliver the power necessary to lift dozens of riders twenty feet up in less than two seconds. The only way this ride can work is with some kind of spring mechanism that stores up energy as the riders come down, then delivers it back to lift them up again.

△ Sure enough, if you walk around to the back you find the springs: large air tanks. When the ride first starts up, the passengers are lifted slowly up into the air. At this point a motor is doing the work of lifting the riders, and it can only do that at a limited rate. But once they are up in the air, they are in possession of enough potential energy to get back to that height the next time. As they come down, a large-diameter piston in a cylinder in the vertical tower compresses air into these tanks, storing the potential energy in the form of air pressure. When they reach the bottom, the compressed air vaults the riders back up again to repeat the nauseating cycle. (Think of it like a basketball bouncing: when the ball hits the ground, it momentarily compresses, storing energy in the air inside.)

▷ Look around on most carnival rides and you'll find a detailed data panel describing the physical dimensions of the ride, the clearance required around it, and its electrical power needs. The bouncy Downdraft ride, for example, uses 115 kilowatts of 208-volt, three-phase power. That's enough to supply the average power usage of about 100 homes, and three times the *maximum* power that a typical 200 amp, 240-volt home electric hookup can supply. That sounds like a lot, but on the other hand, it's equivalent to only about 150 horsepower, like the engine you might find in a midsize car.

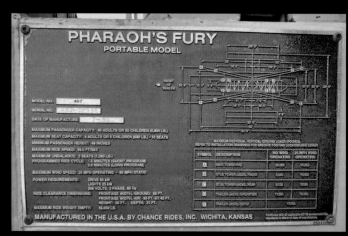

△ This item lists 25 kilowatts as the power required just for the decorative lights on the ride. That's another sign of how old these rides are: a modern one would have LEDs that use a small fraction of the power, and never burn out.

▽ While electric motors dominate on the midway, gasoline and diesel engines have their place as well. The loudest event *by far*, louder even than the country music act, is the tractor pull, in which unreasonably overpowered tractors try to pull a heavy sled as far as they can down a dirt track. They sell earplugs along with the tickets.

The sled has wheels in the back, and a flat plate like a wide ski in the front. A heavy weight starts out at the back end, centered over the wheels. As the pull progresses, the weight moves steadily toward the front, shifting its weight more and more off the wheels and onto the sliding ski, thus making the sled harder and harder to pull. Eventually the tractor can go no farther and the judges measure how far it was able to pull the sled.

◁ This class of competitors is sort of a hybrid between a tractor and a drag racer with massive rear tires and a large engine mounted at the front. You won't see many tractors that look like this plowing the fields.

△ The Mall of America's indoor carnival shows just how quiet even large complex machinery can be when it's powered entirely by electric motors.

△ Grandmothers don't have to scream on the inside.

The gentle sounds of three-phase motors can lull a child to sleep.

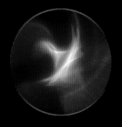

Beyond the Engine

ENGINES AND MOTORS are sometimes called "prime movers" because they are the first—the prime—step in the conversion of energy into some desirable mechanical movement. They are what make the leap from steam pressure, burning fuel, or electric current to physical movement.

Nearly every engine we've looked at so far creates rotary motion—a spinning shaft—as its output. This is not surprising. There are two unique things about circular motion that make it so popular. First, the uniform circular motion of a balanced wheel or shaft is the only form of motion in which all the parts are always moving at a constant speed. If you want a machine that just sits there humming along (or rather, not humming very much), rotary motion is the best bet. Second, it's the only form of motion that doesn't move. That is to say, the outer edge of a round rotating object stays in exactly

the same place, which means it can be *held* in that place using a bearing of some sort.

But as nice as circular motion is, there are plenty of cases where another kind of movement is wanted. For straight-line movement, there are a variety of linear motors that create it directly. For everything else, there are mechanisms. So many mechanisms. Really, you have no idea how many clever mechanisms have been invented to turn one kind of movement into another. A classic work on the subject puts a number on it: *507 Mechanical Movements: Mechanisms and Devices*, by Henry T. Brown, published in 1868. But that's really just the tip of the iceberg. The author assures us that he's left out many types of mechanical movements that he thought were too special-purpose to be of general interest, and each example he did include represents a class of similar designs.

▷ A ball bearing can hold a spinning shaft firmly in place, transmitting mechanical energy without the shaft as a whole going anywhere. This is a tremendously useful property of rotation as a form of motion. A properly balanced precision-machined spinning shaft can appear motionless: it makes no sound or vibration, and if the surface is polished to a mirror finish, you literally cannot see that it's moving.

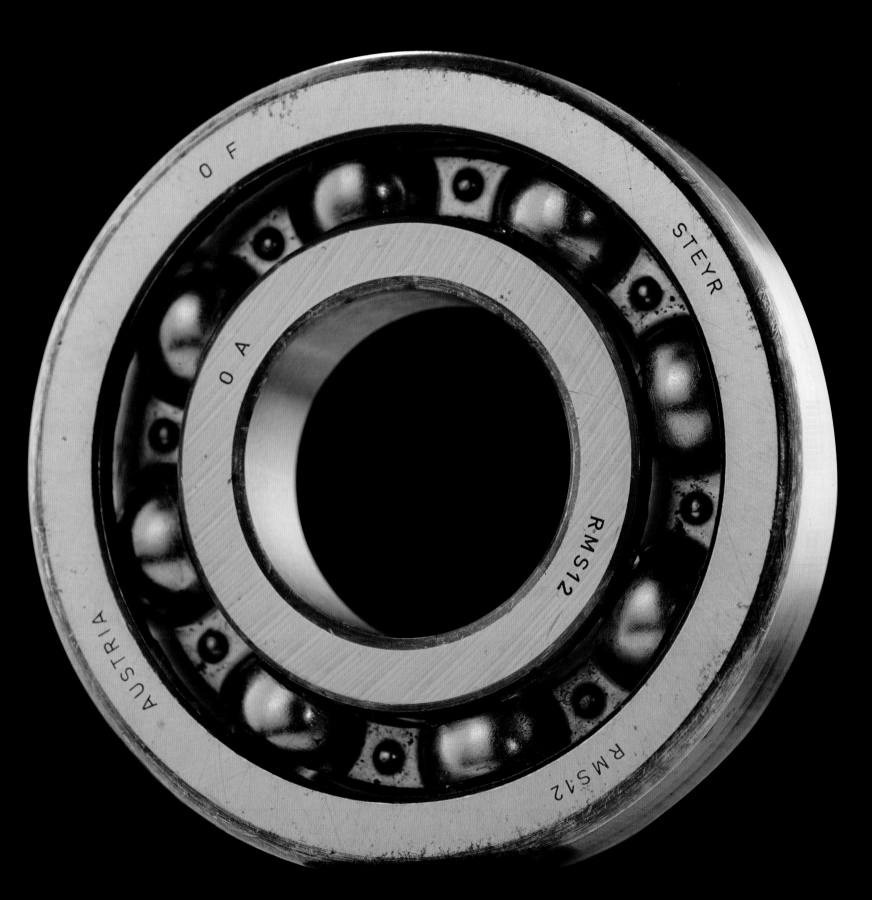

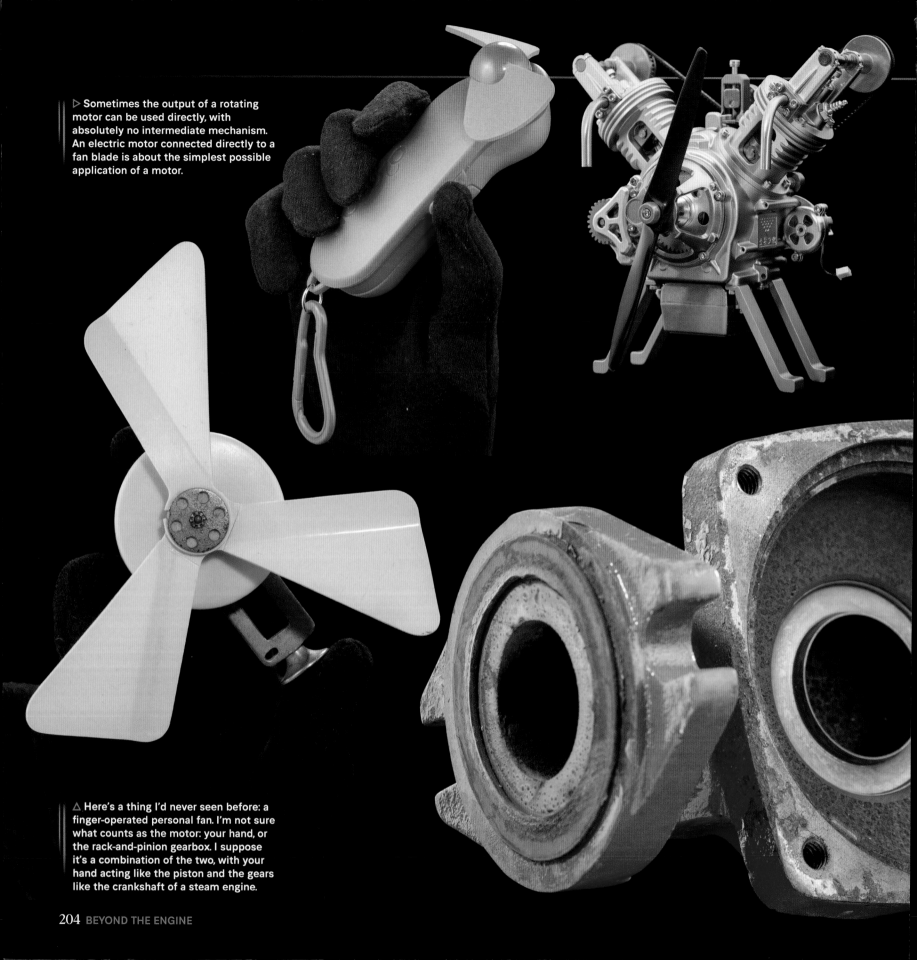

▷ Sometimes the output of a rotating motor can be used directly, with absolutely no intermediate mechanism. An electric motor connected directly to a fan blade is about the simplest possible application of a motor.

△ Here's a thing I'd never seen before: a finger-operated personal fan. I'm not sure what counts as the motor: your hand, or the rack-and-pinion gearbox. I suppose it's a combination of the two, with your hand acting like the piston and the gears like the crankshaft of a steam engine.

◁ In airplanes, real and toy, propellers are often connected directly to the output shaft. Introducing a complex gearbox into an airplane engine causes maintenance problems. The amount of power that needs to be transmitted to a propeller is just too great, and it's easier to control thrust by adjusting the angle of the propeller blades than by changing the speed of the propeller or engine.

▽ Pumps like this one are basically just fans for water. Here, again, there is a direct link between the motor shaft and the moving part, called the impeller. This pump came from the underfloor heating system at my farm. It started screaming and then failed just when I was looking for a pump to photograph for this book.

Vibration

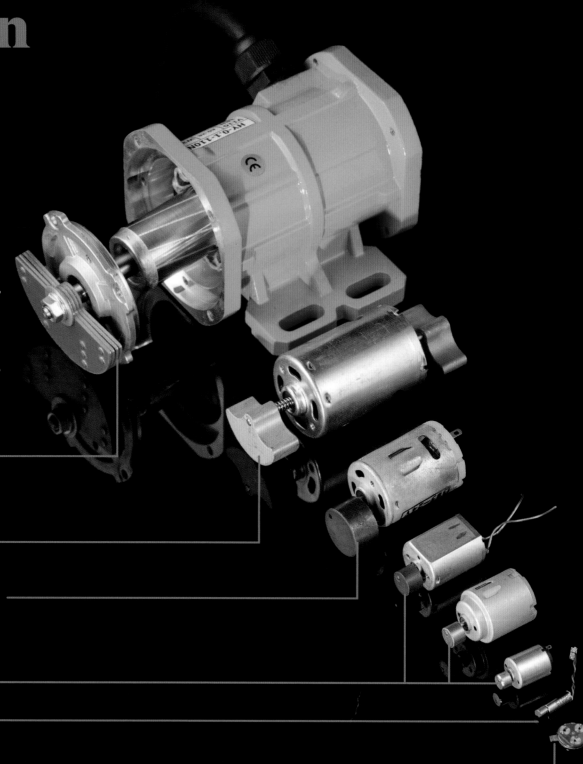

THERE ARE PLENTY of perfectly wholesome applications that generate a little bit—or a lot—of vibration. No gears are needed, just an off-center weight that shakes the motor, and anything it's mounted to, when it spins. Examples range from the size of a coin to the largest I could find, a 40-ton (36 metric ton) seismic vibrator used to map out underground geological structures by sending vibrations through the earth. There may be larger ones out there, but searching for "world's largest vibrator" on the internet is a risky click.

Smaller vibration motors all work pretty much the same way: an ordinary electric motor spins a weight, with the speed, weight, and degree of off-centeredness (the eccentricity) depending on the application.

This one is designed to be bolted to the side of a box, for example to shake sand through a sieve or drive bubbles out of concrete.

This one is from a vibratory tile setter.

This example has an absurdly large weight, way off center. I'm not sure how it avoids bending the shaft unless it's run only at slow speed.

No comment.

This tiny example is typical of the kind used in older cell phones.

This interesting button buzzer is from a modern cell phone.

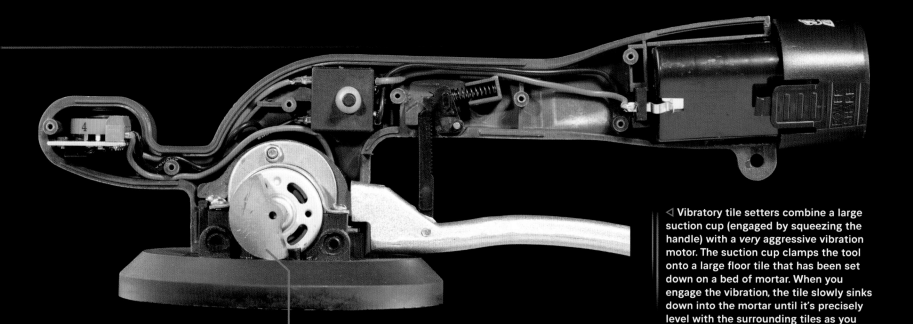

Eccentric weight

◁ Vibratory tile setters combine a large suction cup (engaged by squeezing the handle) with a *very* aggressive vibration motor. The suction cup clamps the tool onto a large floor tile that has been set down on a bed of mortar. When you engage the vibration, the tile slowly sinks down into the mortar until it's precisely level with the surrounding tiles as you install them one by one.

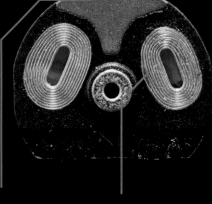

Permanent magnet   Commutator brushes

Spring contacts that connect to the phone circuitry

Off-centered tungsten weight that creates vibration when the rotor is spinning

Coils in the rotor

Commutator contacts on the back side of the disk-shaped rotor

△ This DC motor is a beautiful example of mechanical miniaturization, for the way it folds multiple functions into a minimalist design. It's only about $3/8$ inch (9 mm) across and less than $1/8$ inch (3 mm) thick, yet it's a complete motor with stationary permanent magnets, and a commutator that switches the current in a pair of coils in the rotor. The eccentric (off-center) weight is built right into the rotor, which skips one of the three coils typically found in this type of motor to make room for the weight. My favorite fact about these ultra-miniature vibrators is that they use the element tungsten in their counterweights, because it is the most dense reasonably priced metal available. This lets the thing do a lot more shaking than would be possible with, say, a lead counterweight, which would be just over half the weight of the same volume of tungsten.

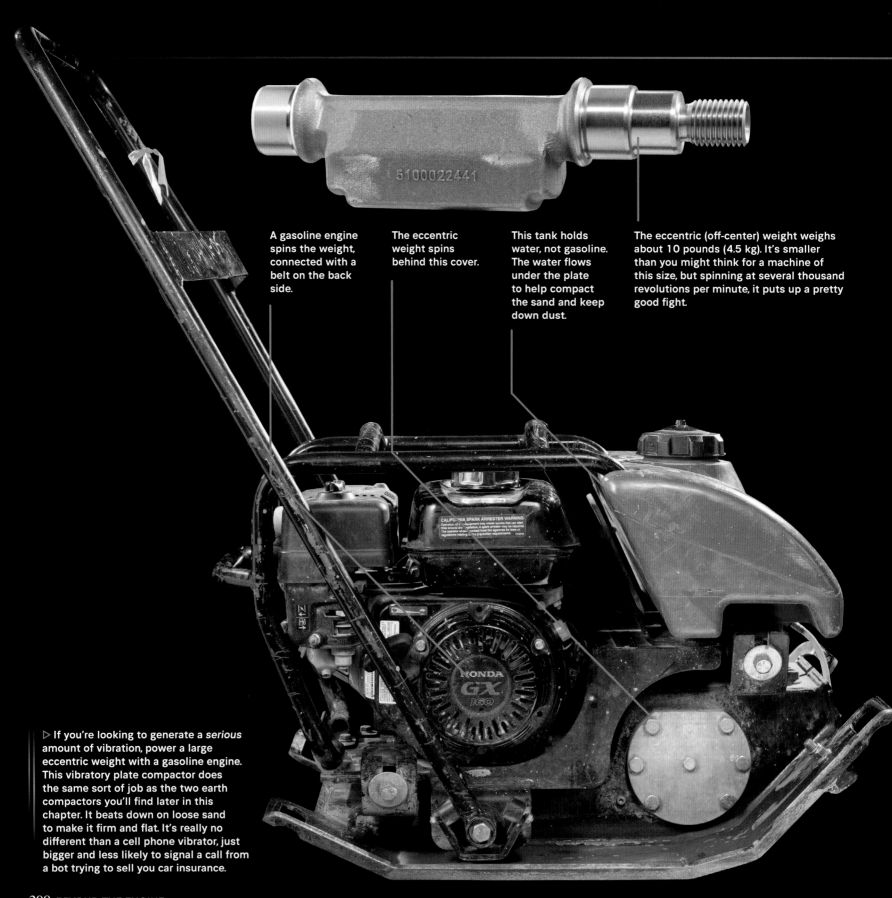

A gasoline engine spins the weight, connected with a belt on the back side.

The eccentric weight spins behind this cover.

This tank holds water, not gasoline. The water flows under the plate to help compact the sand and keep down dust.

The eccentric (off-center) weight weighs about 10 pounds (4.5 kg). It's smaller than you might think for a machine of this size, but spinning at several thousand revolutions per minute, it puts up a pretty good fight.

▷ If you're looking to generate a *serious* amount of vibration, power a large eccentric weight with a gasoline engine. This vibratory plate compactor does the same sort of job as the two earth compactors you'll find later in this chapter. It beats down on loose sand to make it firm and flat. It's really no different than a cell phone vibrator, just bigger and less likely to signal a call from a bot trying to sell you car insurance.

Linear Motors

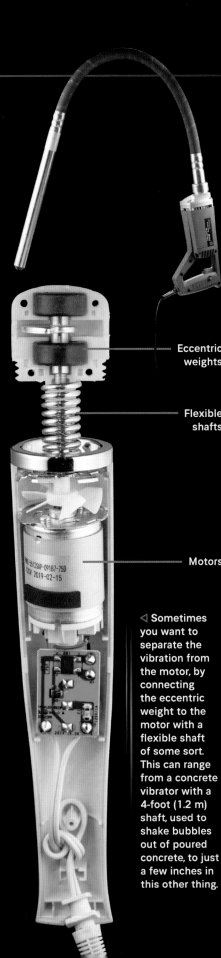

Eccentric weights

Flexible shafts

Motors

◁ Sometimes you want to separate the vibration from the motor, by connecting the eccentric weight to the motor with a flexible shaft of some sort. This can range from a concrete vibrator with a 4-foot (1.2 m) shaft, used to shake bubbles out of poured concrete, to just a few inches in this other thing.

AS WE LEARNED way back at the beginning of chapter 1, nearly all rotating engines actually start with linear motion, like a piston going back and forth. In some cases that linear motion is actually the end goal. So, for each type of rotating engine we've looked at in this book—steam, internal combustion, electric, and hydraulic—there is a linear motor variation. There are even linear motors based on melting wax!

We'll start with steam engines and work our way straight through the other linear motor types.

▽ Taking off in an airplane from the deck of a boat—even a really big boat—is not easy. Fighter planes are designed with small wings that allow them to go fast, but this causes them to have very little lift at slow speeds. Even with their tremendously powerful engines, there's no way for small fighter planes to get going fast enough to take off under their own power before tipping off the side of the boat straight into the ocean. Instead, they have to be thrown off the boat with great force, using what's called a steam catapult.

This is steam from the catapult that was just used.

Underneath this slit in the deck sits a very long steam piston, basically a one-stroke, linear steam engine. When the plane is ready to take off, steam pressure shoots a hook across the deck with tremendous force, dragging the plane up to the necessary speed in just a couple of seconds.

Linear Steam Engines

▷ It's not exactly a steam engine, but this handheld power file has an oscillating linear piston air motor. (It could probably run on steam, but the moisture would cause it to rust very quickly.) The three parts shown nest one inside the other. When air pressure is applied, the inner shuttle works much like the spool valve in a steam engine to alternately apply pressure to one side and then the other. In a way it's acting like both a piston and a valve at the same time, which makes it oscillate back and forth. As it reaches the end of its journey in one direction, it hits the outer sleeve (which is holding the file), causing the sleeve and file to jerk back in the other direction.

Hydraulic cylinder

Ratchet arm

Wheels rotated 90 degrees from their operating position

▷ True hydraulic piston motors, which operate on pressurized water instead of pressurized steam or oil, can sometimes be found on these center pivot irrigation systems. These systems are responsible for the "crop circles" you see when flying over large portions of the American West. Water is pumped from a well in the center out through a long (up to $1/4$ mile/0.4 km!) beam that slowly rotates around the field, covering a circle in water.

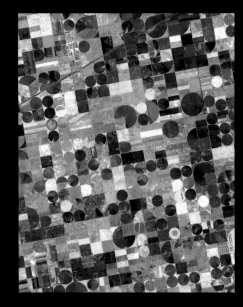

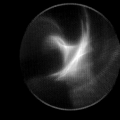

△ At multiple positions along the beam, a hydraulic cylinder draws water from the same pipe that is supplying the spray nozzles. A toggle valve makes the cylinder go up and down in a cycle, and with each cycle a ratchet arm pushes on the teeth of the drive wheels. (This one is currently set up with the wheels rotated 90 degrees from their normal position, so imagine them turned to be in line with each other, moving the pipe sideways.)

Looking at the long beam with multiple drive wheels you may wonder what keeps it going in a straight line. The answer is that the wheels must all turn at slightly different rates, perfectly in sync. In old-school systems this is done with a set of steel cables that stretch the whole length of the beam and connect to control valves at each drive position. If one set of wheels gets a bit ahead of the others, the cables pull its control valve slightly back, slowing it down until the others catch up.

Modern pivot irritation systems use electric motors and electronic guidance systems that let them do fancy things, like reverse direction or irrigate different sections of the field at different rates appropriate to the crop or ground condition at each point, guided by GPS and satellite imagery.

△ The pivot irrigation system is unusual because it has a piston hydraulic motor, but also because its hydraulics actually run on water. The root of the word "hydraulic," *hydro*, comes from the Greek word for water, and is found in words like "hydrate" (get water), "dehydrated" (not enough water), and "r/HydroHomies" (a subreddit for people who are fans of drinking water). But in reality, nearly all hydraulic systems use hydraulic oil, which is similar to a lightweight motor oil. Water would cause rust,

Linear Internal Combustion Engines

▽ This is an earth compactor, used to pound down dirt, sand, or gravel to make a firm surface for pouring concrete or laying bricks. It uses a single-cylinder, linear diesel engine to basically shoot the ground repeatedly. Once it gets going, its own weight falling back down is enough to compress and ignite its diesel fuel, causing it to hop back up again and start a new cycle. I've never seen one of these in action, but the videos make it look pretty exciting to hold on to!

▽ As if a diesel-powered thumper wasn't a crazy enough idea, this is the same linear piston engine concept applied to pogo sticks. A mounting toll of injuries eventually forced the manufacturer to discontinue the product.

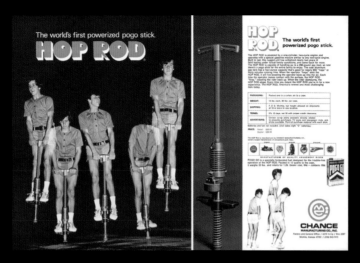

△ There is a small fan that operates *inside* the combustion chamber. When the fuel is ignited, the explosion goes off all around the fan. There's no way this mechanism would be practical in, say, a car engine, where there is an explosion going off dozens of times per second for hours uninterrupted. But in a nail gun it's more like one every second for a short burst, with pauses in between when the whole thing can cool down. The fan presumably helps mix the fuel efficiently and helps clear out the cylinder after the cycle is finished.

△ Now *this* is a great application for the linear piston engine design. I've had this gas-powered nail gun for many years and driven countless thousands of nails with it. It has an internal combustion cylinder that runs on butane fuel. When you press the nose down onto a board, a puff of butane is injected into the cylinder. Then when you pull the trigger, a spark plug ignites the fuel and drives the nail down hard. Think about how hard you can hit a nail with a hammer, and even then you're unlikely to get it all the way down in one go. This tool will drive a 4-inch (10 cm) nail all the way down in one hit every time.

▽ Nail guns need to operate in any orientation, including completely upside down for nailing things to the ceiling. Compressed butane fuel is a liquid, which is good because a lot of it can be stored in a relatively low-pressure can, but also bad because it means you can't have a simple tap at the top or bottom of the can to draw off the liquid. (Think of a can of spray paint that only works when it's mostly right side up.) The solution is to put the liquid butane inside a plastic bag, surrounded by pressurized gas, all inside an outer metal can. The liquid fuel is squeezed out reliably regardless of angle, and without contamination from any propellant gas. (This same idea is used in, for example, spray-on sunscreen that works at any angle.)

Linear Electric Motors

△▽ This vibratory pump uses an electric linear motor to push a piston back and forth, causing one-way valves on the sides to open and close in a way that moves water steadily in one direction.

△ This massage therapy device uses an incredibly simple linear motor. It's nothing more than a coil of wire connected directly to 120-volt AC power from the wall. The oscillating magnetic field created by this coil pulls on the metal plate with a force that varies with the alternating voltage, causing the plate to vibrate strongly. The design is very reminiscent of the relays we looked at in chapter 3, the main difference being that the spring holding the plate away from the coil is much stronger, so the plate never actually snaps down to press against the coil, as it does in a relay. As a testament to how simple, reliable, and indestructible this mechanism is, consider that it's still working perfectly despite the fact that its two-year manufacturer's warranty expired in 1959.

If this were an ordinary electric motor, this part would spin. Instead, it pistons in and out of the black cylinder housing, pushed and pulled by the oscillating magnetic field created by the two wire coils.

▽ This is the linear motor equivalent of the rotary cell phone vibration mechanisms we saw earlier. Apple calls it their Taptic Engine, because Apple likes to make up names for things to make them seem important. Inside, it's basically a solenoid whose only function is to move its plunger back and forth at high speed to create vibration. The geometry is unusual in that the coils are flattened out on either side of the core, but that's just to make it fit in a thin phone.

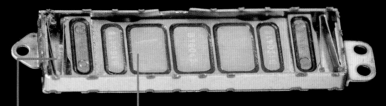

These springs keep the sliding magnets centered.

This set of permanent magnets interacts with the magnetic fields from the flat coils to shift the magnets from left to right.

△ Flat coils, four on each side of the sliding magnets, generate pulsed magnetic fields that make the magnets slide back and forth, creating the sensation of clicks or vibration.

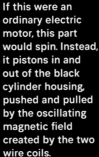

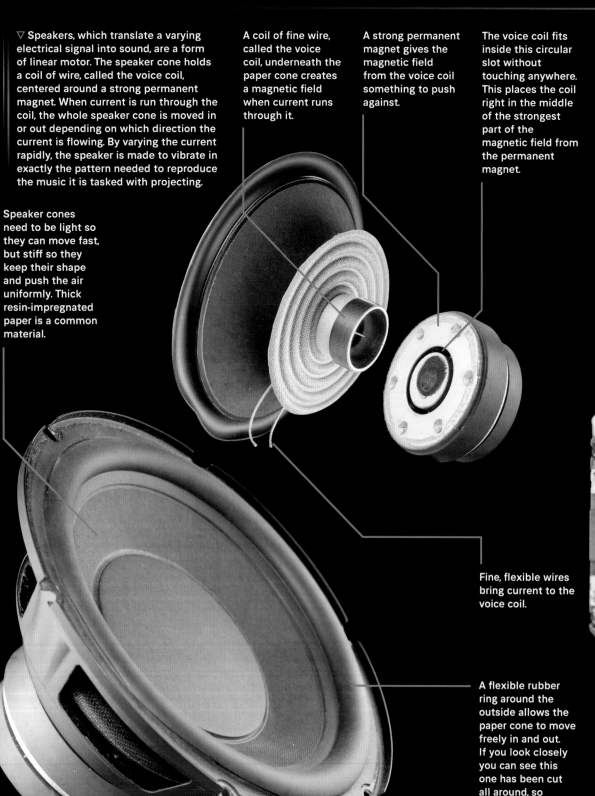

▽ Speakers, which translate a varying electrical signal into sound, are a form of linear motor. The speaker cone holds a coil of wire, called the voice coil, centered around a strong permanent magnet. When current is run through the coil, the whole speaker cone is moved in or out depending on which direction the current is flowing. By varying the current rapidly, the speaker is made to vibrate in exactly the pattern needed to reproduce the music it is tasked with projecting.

Speaker cones need to be light so they can move fast, but stiff so they keep their shape and push the air uniformly. Thick resin-impregnated paper is a common material.

A coil of fine wire, called the voice coil, underneath the paper cone creates a magnetic field when current runs through it.

A strong permanent magnet gives the magnetic field from the voice coil something to push against.

The voice coil fits inside this circular slot without touching anywhere. This places the coil right in the middle of the strongest part of the magnetic field from the permanent magnet.

▷ From the side we see tiny wires that both conduct electrical signals to the stabilization and focus motors, and act as springs supporting and centering the movable parts (in this picture the lens is shifted slightly down under its own weight).

▽ If you like the miniaturization of the cell phone vibrator, you'll love this thing! It's about the same size as the vibrator ($1/2$ inch/12 mm square and $1/8$ inch/3 mm thick), but it is a complete camera image stabilization *and* focusing mechanism in one. It includes *three* independent linear motors. They are called voice coil actuators because they work basically the same way as the voice coil in the speaker: coils of wire push and pull magnetically against permanent magnets.

Fine, flexible wires bring current to the voice coil.

A flexible rubber ring around the outside allows the paper cone to move freely in and out. If you look closely you can see this one has been cut all around, so I could take the cone out.

This whole unit can move left/ right and up/down to shift the lens, keeping the image centered on the image sensor as the phone moves slightly.

The lens can independently move front to back, focusing the image.

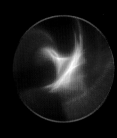

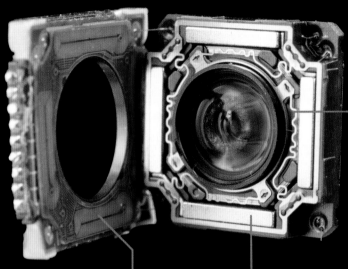

These squiggly wires are springs, and electrical conductors, that support the lens and bring electric current to the focus coil.

▽ Underneath the mechanism we just learned about sits the true champion of miniaturization: the 12-megapixel image sensor. The iridescent colors are created by diffraction of light by the 36 million individual light sensors (red, green, and blue for each of the 12 million image pixels).

Hair-thin spring wires

△ With the spring wires cut on one side, we can open the unit up.

These coils on the stationary base push and pull on the permanent magnets, shifting the stabilization mechanism up/down or left/right.

Four strong permanent magnets work together with both the stabilization coils in the base, and the focus voice coil that we haven't seen yet.

Here is a better view of the stunningly beautiful stabilization voice coils.

▽ With the lens module removed, we can see the focus voice coil. It pushes and pulls on the same permanent magnets as the stabilization coils, but in the other direction.

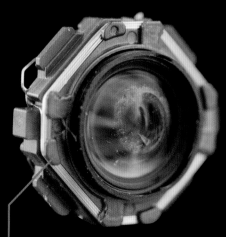

Focus voice coil

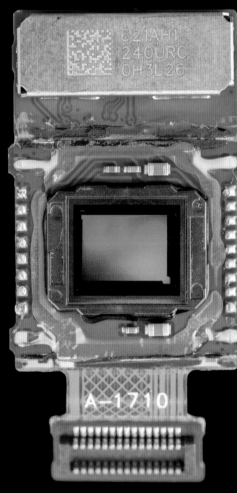

▷ Maglev (magnetic levitation) trains are not common, but there are some in regular commercial service, including this one in Shanghai. They are propelled by linear induction motors, which are basically the same idea as a standard induction motor, unrolled into a long, straight track.

▽ The more extreme version of a linear motor is represented by this rail gun. It accelerates metal projectiles to over 2 miles per second (3 km/sec) using magnetic forces alone. Rail gun projectiles typically don't contain any explosive: when they hit a target, they blow it to bits by kinetic energy (the energy of a fast-moving object) alone.

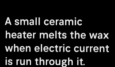

A small ceramic heater melts the wax when electric current is run through it.

▷ Rail guns are the fastest linear motor, and this is the slowest. It's what's commonly called a "wax motor," though it's really more of a linear actuator than a motor in the traditional sense. The motor part of the name may be questionable, but the wax part is to be taken literally. Inside the housing, there is a small cylinder filled with a wax that's solid at room temperature. When an electric current is run through the two contacts, the wax heats up and melts, which causes it to expand enough to push the black rod a good distance out. And when I say push, I mean it pushes *hard*. Phase changes (from solid to liquid or liquid to solid) can exert a huge amount of force. Witness, for example, the way that water freezing into ice can crack rocks and concrete with ease. This is the same idea in reverse and is deployed to, for example, temporarily hold the lid of a washing machine closed so it can't be opened while the machine is running. When the current is switched off, the wax cools in a few seconds, and a strong spring inside recompresses it, withdrawing the rod slowly. The fact that this happens in response to the power being cut is why this type of device is used in safety interlocks: even if the machine is completely shut down or unplugged, you won't be able to open the door for a few seconds while it spins down and the wax cools, but then you *will* be able to open the door, even though there's no power left to actuate any sort of release mechanism.

Inside the plastic housing is a sturdy metal cylinder containing a small amount of wax.

▷ When the current is shut off, the wax cools and the piston is returned to its off position by a *powerful* spring. Seriously, if you ever take one of these apart, watch out for the spring.

△ Here's a clever application of a much larger wax motor: a piston that automatically opens the windows in a greenhouse when the temperature gets too high. The rod extends about 4 inches (10 cm) and pushes so hard that it can easily open a heavy window. The weight of the window automatically pushes the rod back in when the temperature goes down. No electronics, no control system, no power supply—it just opens and closes windows all by itself, using the thermal energy of the air around it to drive the action.

Converting Motions

WE'VE JUST SEEN examples of motors that directly create linear motion, but often the most practical solution is to use an ordinary rotating engine and add some kind of mechanism to convert its rotary motion into linear motion.

We've seen an earth compactor that thumps the ground directly with a linear piston. But this obviously sensible design is actually a rare and exotic curiosity. The compactors regularly used on construction sites have a perfectly ordinary rotating gasoline engine, which drives a set of gears followed by a crank arm that pushes the bottom half of the machine up and down, causing it to thump the ground really hard.

If you think about it, this might seem a bit nuts: You have a piston in the engine going back and forth, turning the crank on the crankshaft, which turns some gears, which turns another crank to make the "piston" on the bottom go back and forth. Why not cut out the middle mechanism and make the piston thump the ground directly?

The answer hinges largely on a mismatch between the rate at which a gasoline engine likes to operate (a dozen strokes per second or more) and the rate at which a worker may want to pound the ground (more like once a second). The gears solve that problem.

It's also partly due to mass-manufacturing efficiency. Making a reliable internal combustion engine is *not easy*. The precision fit required between the piston and cylinder can only be achieved with highly specialized precision machinery, and the cost of doing that is reasonable only when you're making literally millions of them. There are only a handful of companies in the world that actually make engines; all the tool manufacturers buy their engines as stock or semi-customized components from these companies. The total world market for earth compactors just isn't large enough to support the production of a linear engine version that's cost-competitive with one that uses an off-the-shelf gasoline engine, even though it's mechanically far more complicated.

And then there's the issue of maintenance: Because the rotating engine is a standard model used in many other power tools, it's much easier to find parts for repair, not to mention someone who knows how to fix them. A unique engine used only on earth compactors? Good luck getting that fixed.

Whatever the reason, the fact is that the great majority of complicated motions are created by rotating motors plus some kind of mechanical linkage.

△ I call these things thumpers, because that's what they do: They go thump-thump-thump, pounding the ground as hard as they can. This one has an ordinary four-stroke engine and a mechanism that converts its rotary motion into the desired thumping action.

▷ I've removed the cover so you can see the thumping mechanism. It's a simple but beautifully made gear reduction chain followed by a crank and connecting rod that makes the bottom half of the machine push in and out. The weight of the machine is thus bounced up and down on the ground.

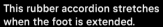

This rubber accordion stretches when the foot is extended.

Armatron

WE'VE JUST SEEN one example of a simple mechanical linkage: a way to turn circular motion into straight line motion. As I mentioned at the start of this chapter, there are thousands of such mechanisms that offer any conceivable form of non-rotary motion. Whole books have been dedicated to the subject, and the internet is full of fascinating animations. The subject is in some ways even bigger than the subject of engines. So rather than try to list and categorize a lot of different kinds of linkages, I'm going to skip right ahead to some really fun examples.

By far my favorite mechanical linkage is the Armatron toy robotic arm. It's a dizzyingly complex and sophisticated relic of the 1980s. This thing has six fully independent degrees of freedom:

1 The arm as a whole rotates.

2 The shoulder joint goes up and down.

3 The elbow joint goes up and down.

4 The wrist bends left and right.

5 The wrist rotates.

6 The hand opens and closes.

▷ This unassuming plaything is one of the most mechanically intricate toys of its time. From the outside it looks like nothing special, but inside there are more gears of more kinds than I had imagined possible when I first took one apart. And it took a while for me to realize the subtle genius involved in the way it changes gears. Whether this genius is strictly necessary is another question. It seems possible that a much simpler, more direct gearbox design could have worked, but faced with the obvious brilliance of what we see here, I tend to think that if there was a better way, the people making this thing would have found it.

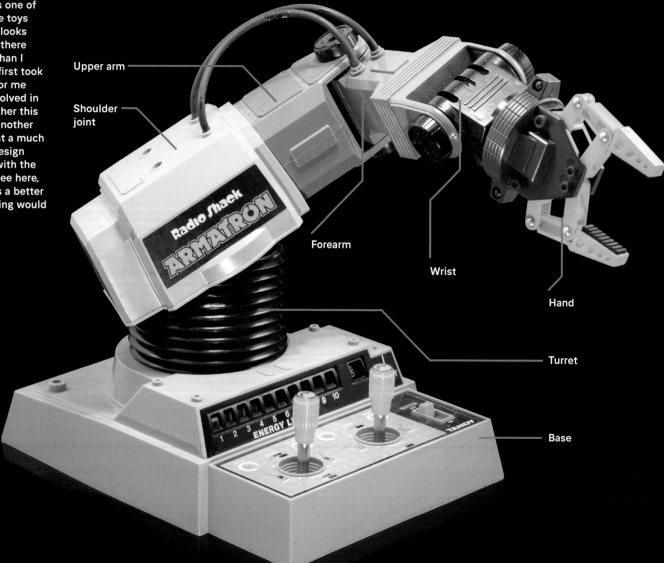

Upper arm

Shoulder joint

Radio Shack ARMATRON

Forearm

Wrist

Hand

Turret

Base

Mind-blowingly, the Armatron has only *one motor*, a little electric one down in the base. Everything else operates with mechanical linkages, including the two joysticks, each of which can be moved left/right or up/down, as well as rotated clockwise or counterclockwise. It's a simple motor connected to a fantastically complicated transmission/gearbox assembly that delivers the rotary power of the motor down one of six different gear chains, from the base all the way out into the hand.

Modern versions of this toy use individual electric motors at each joint, which allows for much greater flexibility. With shifts in the relative costs of designing and making mechanical parts versus electrical ones, it's also cheaper. But it's not nearly as impressive a design and engineering feat.

In the base of the turret we see five pinion gears, each of which engages with one of the internal-tooth gears in the base.

This turret is the first part of the arm that moves.

▷ Each of the six mechanical axes of motion is carried through the arm by a separate gear chain. All of them start with gears in the base of the turret and are brought up through the joints with separate shafts and gears.

This pit of internal-tooth gears looks alarmingly like that sand creature pit in the original *Star Wars*, doesn't it? No fewer than *five* independent gears are stacked and nested into each other.

▷ With the cover removed we can see the true complexity of this thing. In the base is a set of six rotating gearboxes, each with several internal gears. When one of the joysticks is pushed forward or backward, left or right, or twisted clockwise or counterclockwise, one of the six gearboxes rotates forward or backward, bringing one of its two sets of gears into contact with a gear on the driveshaft, as well as one of the six gear chains that lead up into the arm. This completes a mechanical linkage and sets in motion one of the joints.

▷ This arm replaces simple toy motors with digitally controlled servomotors, which are barely any more expensive. They require a sophisticated electronic control system to run, but that contributes virtually nothing to the cost because it's just a microchip, and no matter how complicated it is, it's not going to cost more than a few cents. Using servos means this arm can be controlled by a computer and programmed to execute complex movements.

Each of these six drums is a separate gearbox, which can rotate forward or backward to engage a different set of gears.

These gears engage the "pit of doom" gears on the bottom of the turret.

This single motor drives all the movements.

▷ The modern version of a toy robot arm uses individual electric motors at each joint. This makes for a much simpler physical design, with few disadvantages over the old mechanical version. Today's electric motors are not only cheap but also very light and powerful, thanks to rare earth magnets. It would make no sense to build a machine like the Armatron anymore.

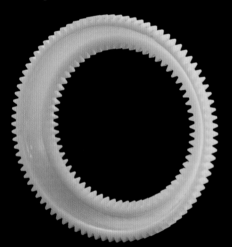

△ Each ring gear in the base has teeth on both the inside and the outside. If a gear like this had to be made of precision-machined metal, it would be very expensive. But as an injection-molded plastic part, it costs a fraction of a penny to make. (The mold to make the plastic part is *very* expensive, but if you're making millions of them this doesn't matter.) Injection-molded plastic is the real secret to why the Armatron made sense with just a single motor: even though there are a huge number of intricately shaped plastic parts, they can all be made incredibly cheaply.

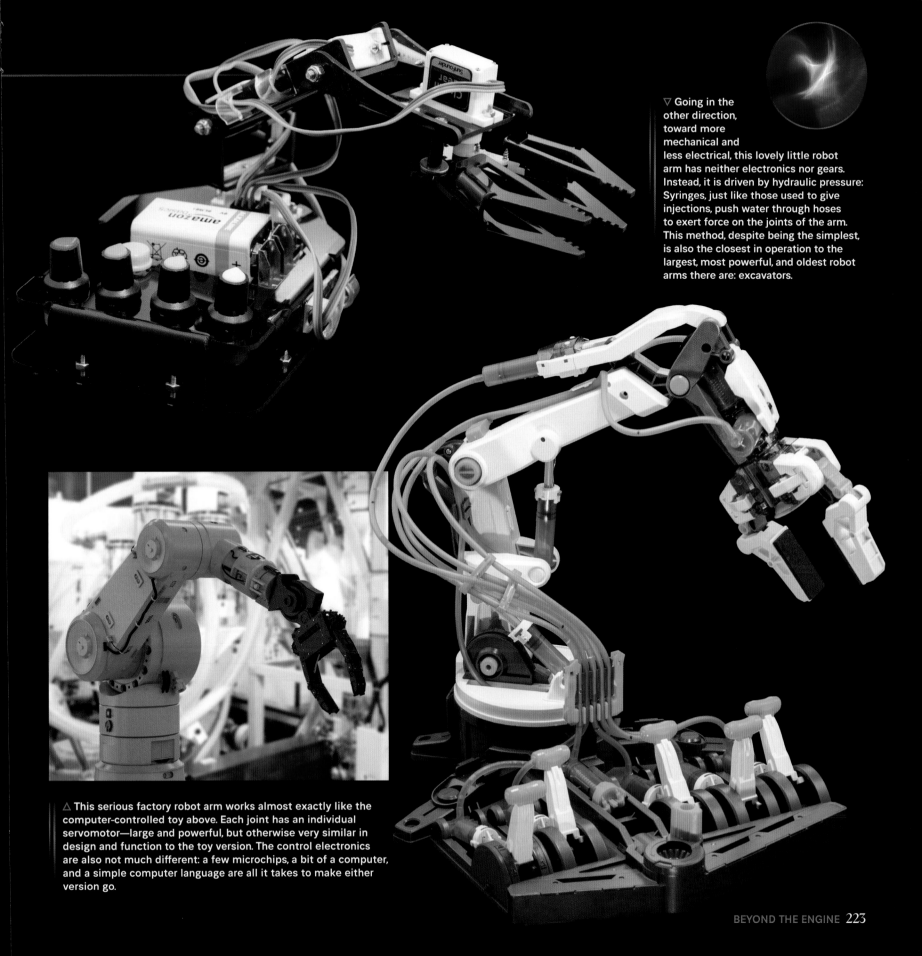

▽ Going in the other direction, toward more mechanical and less electrical, this lovely little robot arm has neither electronics nor gears. Instead, it is driven by hydraulic pressure: Syringes, just like those used to give injections, push water through hoses to exert force on the joints of the arm. This method, despite being the simplest, is also the closest in operation to the largest, most powerful, and oldest robot arms there are: excavators.

△ This serious factory robot arm works almost exactly like the computer-controlled toy above. Each joint has an individual servomotor—large and powerful, but otherwise very similar in design and function to the toy version. The control electronics are also not much different: a few microchips, a bit of a computer, and a simple computer language are all it takes to make either version go.

▷ Really large-scale robot arms are all hydraulically operated, and even today use a power distribution mechanism that is more like the Armatron than like the individual motors of an electric robot arm. Machines like this have a single powerful high-pressure oil pump coupled directly to their main gasoline or diesel engine. The pressurized oil flows into a manifold that distributes it to a series of valves, one for each axis of movement in the machine. This block of valves is the equivalent of the complicated gearbox in the Armatron: It decides where, when, how strongly, and in which direction hydraulic fluid will be sent. Instead of gear linkages to carry the power up into the joints, hoses carry pressurized oil to hydraulic cylinders that push directly on the joints, bending them one way or another.

It turns out that when you need a *lot* of force—say many tons of it—hydraulics can't be beat. The operating pressure in a machine like this is around 3,000 pounds per square inch (3,000 psi, or 20 MPa [megapascals] in metric units). Multiplied by the cross-sectional area inside the cylinders, you can easily generate a force of dozens of tons at any point in the machine where you care to put a cylinder.

These valves, like those found in the cab of any piece of equipment that runs on hydraulics, do the same thing as the gearbox in the Armatron: They direct hydraulic oil to specific joints. Each lever slides a spool valve inside up or down, directing fluid to one or the other of the outlet ports.

Hydraulic fluid from the high-pressure pump enters here.

When the lever below is pushed up, fluid is directed out this port.

When the lever is pushed down, fluid comes out this port.

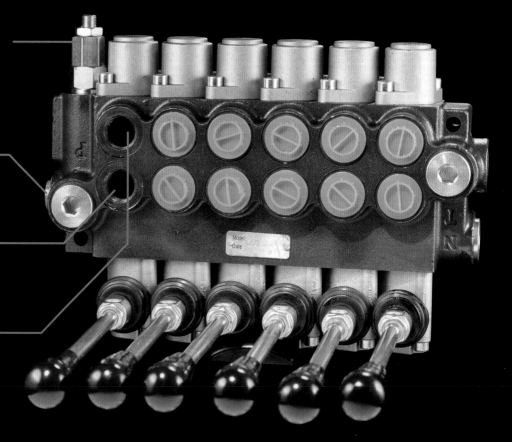

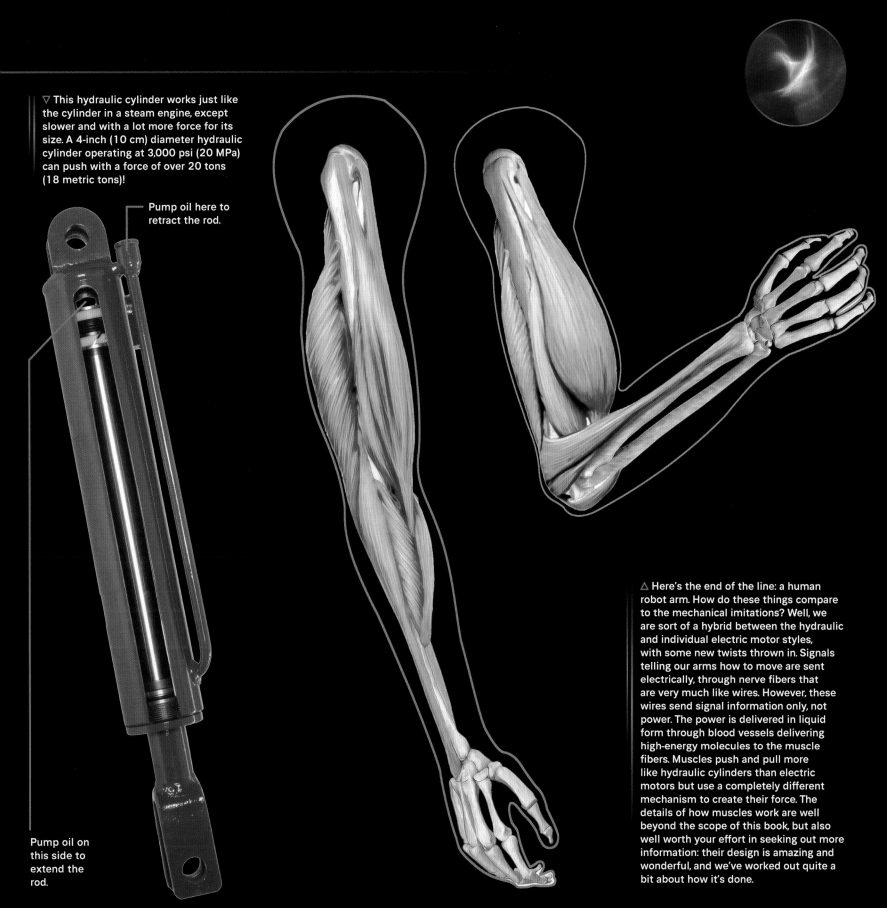

▽ This hydraulic cylinder works just like the cylinder in a steam engine, except slower and with a lot more force for its size. A 4-inch (10 cm) diameter hydraulic cylinder operating at 3,000 psi (20 MPa) can push with a force of over 20 tons (18 metric tons)!

Pump oil here to retract the rod.

Pump oil on this side to extend the rod.

△ Here's the end of the line: a human robot arm. How do these things compare to the mechanical imitations? Well, we are sort of a hybrid between the hydraulic and individual electric motor styles, with some new twists thrown in. Signals telling our arms how to move are sent electrically, through nerve fibers that are very much like wires. However, these wires send signal information only, not power. The power is delivered in liquid form through blood vessels delivering high-energy molecules to the muscle fibers. Muscles push and pull more like hydraulic cylinders than electric motors but use a completely different mechanism to create their force. The details of how muscles work are well beyond the scope of this book, but also well worth your effort in seeking out more information: their design is amazing and wonderful, and we've worked out quite a bit about how it's done.

Mechanical Plotting

PERHAPS THE ULTIMATE example of converting rotary motion into complex side-to-side motion is found in the autopen, a device used to automatically write a signature using a real pen, and pointedly *not* using any form of electronics. Devices like this have been used by world leaders since the time of Thomas Jefferson to execute the thousands of signatures they are required to deliver over their time in office.

The device has two irregularly shaped wheels (cams) that encode the up/down and left/right movements needed to re-create the signature. As they turn slowly, two cam followers ride along the irregular outside edges, moving in and out. These movements are transmitted to the pen through a series of lever arms. The relationship between the cam wheel diameter and the movement of the pen is a bit complicated, and the two movements interact with each other, so it might seem complicated to figure out the required shape of the cam disks. But creating the wheels is actually quite easy: The device is just used in reverse. The user writes their signature in the normal way, using the pen connected to the lever arms with the cam followers replaced by pencils drawing on slowly rotating blank disks. The disks are then cut out along the trace left by the pencil. No matter how complicated the mechanical linkage, this shape will necessarily trace out the same signature when the cam followers are put back in place.

A modern computerized pen-plotter could trivially do the same job, but there's a lot to be said for never encoding the precise strokes of a president's signature in electronic form. No matter how clever hackers may be, they can't download data that exists only in the form of an oddly shaped, closely guarded disk.

▷ This autopen was not used by any president, because it's a model I made to illustrate how they work.

▷ This Autopen Model 50, made by the International Autopen Company, was used by President John F. Kennedy.

Cam followers are pressed against the outside of the disks by springs (rubber hair bands in this model), which ensures that they follow the ins and outs of the disks' edges. The blue follower tracks the blue disk, and the yellow follower tracks the yellow disk.

Two irregularly shaped cams (yellow and blue in this model) control the movement of the pen.

A real autopen would have an electric motor or wind–up mechanism to turn the disks at a steady rate. This model has a hand crank.

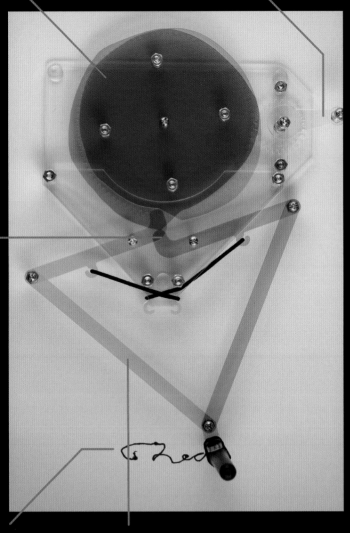

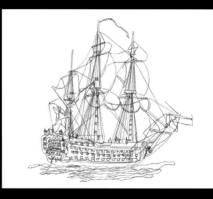

The disks are "programmed" to write my name. OK, it's not very tidy, but they're doing their best!

Two cross-linked lever arms transfer the movement of the cam followers to movement of the pen. (A third disk, not shown in this model, can be used to lift and lower the pen for separate words.)

△ Autopen-type mechanisms can get pretty extreme. This purely mechanical automaton, made in London around the year 1800, has so many cams of such exquisite shapes that it can draw seven different elaborate designs—three poems and four drawings. And just to make things extra complicated, the information in the cams is transmitted through complex linkages to the hands of a puppet, which appears to do the drawings on its own. These devices were considered some of the foremost marvels of their day, for good reason! This one is believed to be the most complex ever made.

▷ An autopen would be a pretty unusual thing to have in your home, but a very similar idea is commonly found in any home sewing machine that's able to do complicated stitches. Making these stitches requires controlling two directions of movement at the same time: the needle moving side-to-side, and the fabric moving front-to-back. A simple zigzag stitch only needs to move the needle left to right while the fabric advances at a steady rate. But if you want to do a fancier curlicue stitch, you need to move the fabric sometimes forward and sometimes backward, at the same time the needle is moving left to right. This is done using two separate complicated cams, just like in an autopen.

Modern sewing machines typically have a dial on the front that lets you choose from a fixed set of fancy stitches. The dial moves a pair of cam followers across a complicated multi-segmented cam that has a pair of unique shapes for each stitch.

△ This 1953 Elna Supermatic sewing machine has a mounting point where you can insert one double cam at a time to select which fancy stitch you want. This is less convenient than just turning a dial, but on the other hand, it means there's no limit to how many different stitches the machine can do. You could even

With the stitch length/direction lever all the way down in the A position, a second cam follower snaps in from the right to ride along the top

When you move this lever anywhere to the right of 0, the metal cam follower just above it snaps to the right, pressing against the bottom half of whatever cam is installed at the time. (In this picture there's no cam present, so the follower goes all the way in to touch the mounting hub.)

△ With a cam in place, both cam followers can do their job riding the contours of the cam disk. It takes eighteen stitches for the cam to make one complete revolution, so you can program quite complicated stitching patterns.

▷ Dozens of different cam sets are available, each marked with a picture of the stitch it will produce.

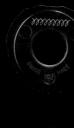

Some of them are quite elaborate, with stitches that go alternately left and right, while going forward and backward at the same time. This is a true programmable autopen in stitching form.

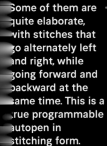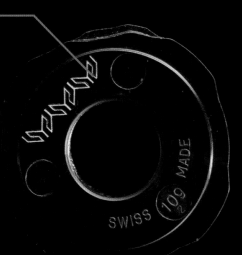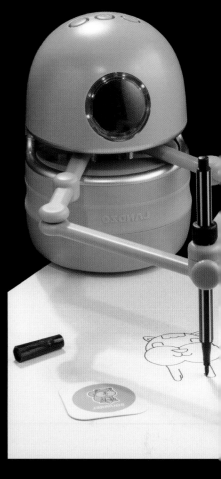

SWISS 109 MADE

◁ This book was written in the middle of the COVID-19 pandemic, during which my girlfriend Maribel and I decided to switch our efforts over to making cloth face masks. Modern home sewing machines are hopeless pieces of trashy plastic, and modern industrial machines are generally inconvenient to keep at home when you're in quarantine. The best machines *by far* to sew at home are sturdy, well-built home models from the 1950s through the 1980s. I started mask-making with a 1980s Singer that used to belong to my mother. When my daughter started making masks too, I gave her the Singer and switched to the astonishingly beautiful 1953 Elna Supermatic shown here. I got it for $25 at an (online) auction, not really knowing much about it. To my delight it turned out to be the smoothest-running sewing machine I've ever seen. It's so quiet you can literally hear the thread being pulled through the fabric and tensioner, a sound I have never heard before in an operating sewing machine.

△ This is the modern (toy) version of an autopen. It uses electric motors and computer control to draw little doodles using two arms to control the two directions of pen movement. It is mechanically far, far more complex than the purely mechanical version. The computer chip alone has thousands of times more "moving parts," and that's not even counting the digital camera and stepper motors. Yet modern manufacturing is so fiercely efficient that this complexity can be delivered in a cheap toy for children.

Lots of a Few or a Few of Lots

THERE IS A SUBTLE but important difference between the older Armatron and the newer computer-controlled robot arm we saw earlier—one that may not be obvious if you aren't a student of how things are made.

The Armatron was only a realistic product to make if the company planning it had a lot of money to invest, and could be sure they could sell a *lot* of them. Because of the large number of complex and expensive injection molds needed to make its many gears and housing parts, it would have cost several hundred thousand dollars to create the tooling necessary to make the first one. No one is going to do that unless they are reasonably sure they can sell a boatload of them. At least part of that calculation is thinking about whether there are other, similar products that might compete with yours. In the case of the Armatron, there was nothing else like it on the market, and the chances of someone coming out with a competitor were slim to none, given how complicated the thing is.

The computer-controlled arm, on the other hand, is made entirely using manufacturing methods that are practical for very small production runs. The custom-shaped parts are all laser-cut and bent acrylic: There are no molds or other upfront costs at all. Custom circuit boards can be

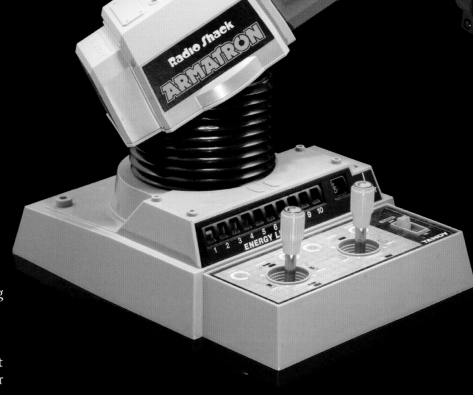

△ The Armatron and the Arduino tell a tale of two worlds.

made to order in highly automated factories, with a minimum order of just a few dozen, or you can use stock microcontroller boards that cost just a dollar or two. The other parts—screws, servomotors, microchips, connectors—are all off-the-shelf parts made in huge quantities by parts suppliers who are selling the same thing to thousands of different customers making thousands of different products.

The Armatron required the resources of a largish company to

pull off, and could only exist in a market where it was mostly alone. The computer-controlled arm could be designed and put on the market by one or two people working in a small shop (I should know, since the methods used to do it are pretty much the same as what I do with my mechanicalgifs.com models).

The key effect of having a low upfront cost is that the modern example can coexist in a market with dozens of similar competing products. That practically defines

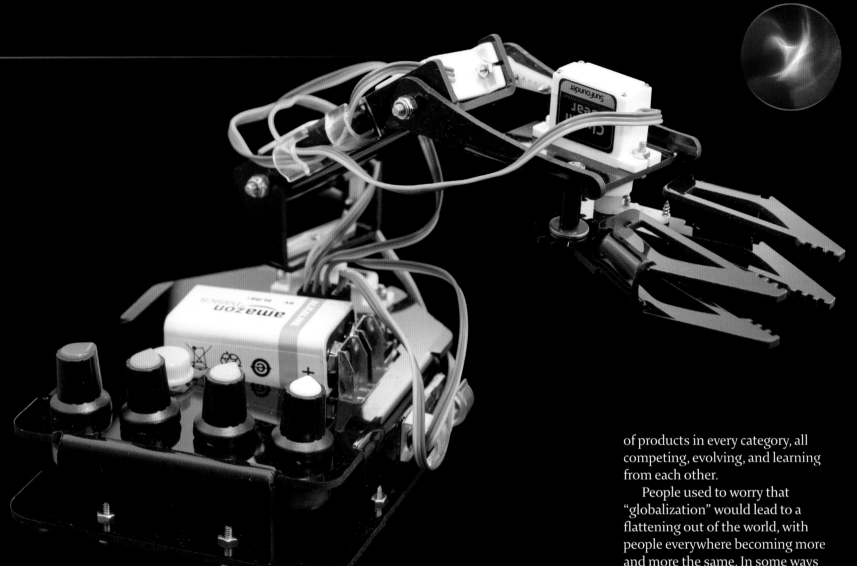

the world today compared to the Armatron's world of forty years ago. We have gone from a small number of products consumed by a large number of people, to a large number of products each with a smaller audience.

That's true for everything from toys to television channels. In the 1980s there were exactly four channels available to the average viewer in the US, and the cost of starting a new network was in the billions. Can you imagine explaining to a network executive of the time that one day there would be so many "channels" that it wasn't even possible to count them, and anyone who wanted to create a new one could do it in a couple of minutes *for free*!

Creating a new toy isn't quite *that* easy, but with laser cutters, 3D printers, and access to wholesale supplies of parts through the internet, it's much, much easier than it used to be. As a result, we have an explosion of creativity and a diversity of products in every category, all competing, evolving, and learning from each other.

People used to worry that "globalization" would lead to a flattening out of the world, with people everywhere becoming more and more the same. In some ways this has happened: McDonald's may be somewhat different in Italy, but it's still McDonald's, and it's everywhere. But in other ways we've seen an explosion in diversity. Incredibly specific virtual communities come together around a shared, highly obscure interest. Products are created and marketed online in a matter of days or weeks instead of months or years. Your town may have lost all its retail stores to a nearby Walmart, but you can have delivered overnight a thousand times as many things as you could ever have gotten in all those local shops put together.

Acknowledgments

THIS BOOK WAS WRITTEN, edited, laid out, and printed almost entirely during the great pandemic that started in 2020 and went on for . . . well, I don't know, because at the time of this writing, we're into 2022 and things are not going well.

I can't honestly say I had a particularly hard time during the pandemic. Not the way many people did. Authoring is a lonely profession, and I don't much like people anyway, so no great loss not being able to go to, I don't know, a party, or whatever people used to do. But it still takes a toll, watching the world burn.

A large part of the reason this book is coming out a year later than scheduled is that for a full year I diverted most of my attention to making masks with my girlfriend Maribel. She designed a superior style of mask, I gathered tools and raw materials, and together we built a business and sold many thousands of super-comfortable masks lined with two layers of pure silk. I have to thank her for making that business possible, but I guess that's not exactly thanking her for helping finish the book? On the other hand, Maribel's son, Toby, contributed more directly by modeling for several of the photos.

So, naturally, I have to thank my publisher, Becky, for putting up with such a delay and generally being a great person to work with over many years.

As for the book itself, I must start by thanking Nick Mann for his outstanding work in taking nearly all the photographs in this book. (You can tell which ones I took: they're not as good.) He has been a steadfast presence for so long, it's hard to imagine working without him. I also need to thank Nick's stepson, Frankie, for being exactly the same size as my friend Donald's huge piston and connecting rod. Thanks also to Gretchen for keeping everything going around the studio and trying to keep me out of accounting jail.

This book involved a larger than usual amount of large things: car engines, an entire car, large power tools, electric motors too heavy for me to lift, and so on. For help with these I thank Bobbie and Koatie, and their children Tristan, Alexus, Quinton, and Brianna—good friends and the caretakers of my farm. Without their help I'm sure I would have dropped something very heavy on my foot.

I am very much in debt to Allen and Nancy Strong for allowing Nick and me to photograph in their amazing antique-car showroom. It is a wonderful thing that such a place exists, and that it's so close to where I live.

I also greatly appreciate the support of Raymond and the others at Four Acre Wood Products in Arthur, Illinois, for allowing me to photograph in their factory, and for frank and enlightening discussions about the Amish approach to technology.

Thanks to the fine folks at the Stephenson County Antique Engine Club for permission to photograph their fine old Cooper Corliss engine.

For acquiring many things, including the outstanding metal engine models shown on the cover of this book, I want to thank the person I affectionately know as Chinese Nina (to distinguish her from Ex-Wife Nina). When I need something from China, she gets it for me, has it shipped to her in Shenzhen (so the seller doesn't know it's for an American buyer and double the price), and then forwards it to me. Everyone should have a Chinese Nina. (And thanks also to Guangzhou Teching Culture Development Co., Ltd. for making those metal engine models.)

In past books I have always thanked my children, Addie, Emma, and Connor, for being generally helpful and fine examples of the species. But now they are all grown up and moved out, so I guess I have to thank them for being fully functional adults without much need for further development work on my part.

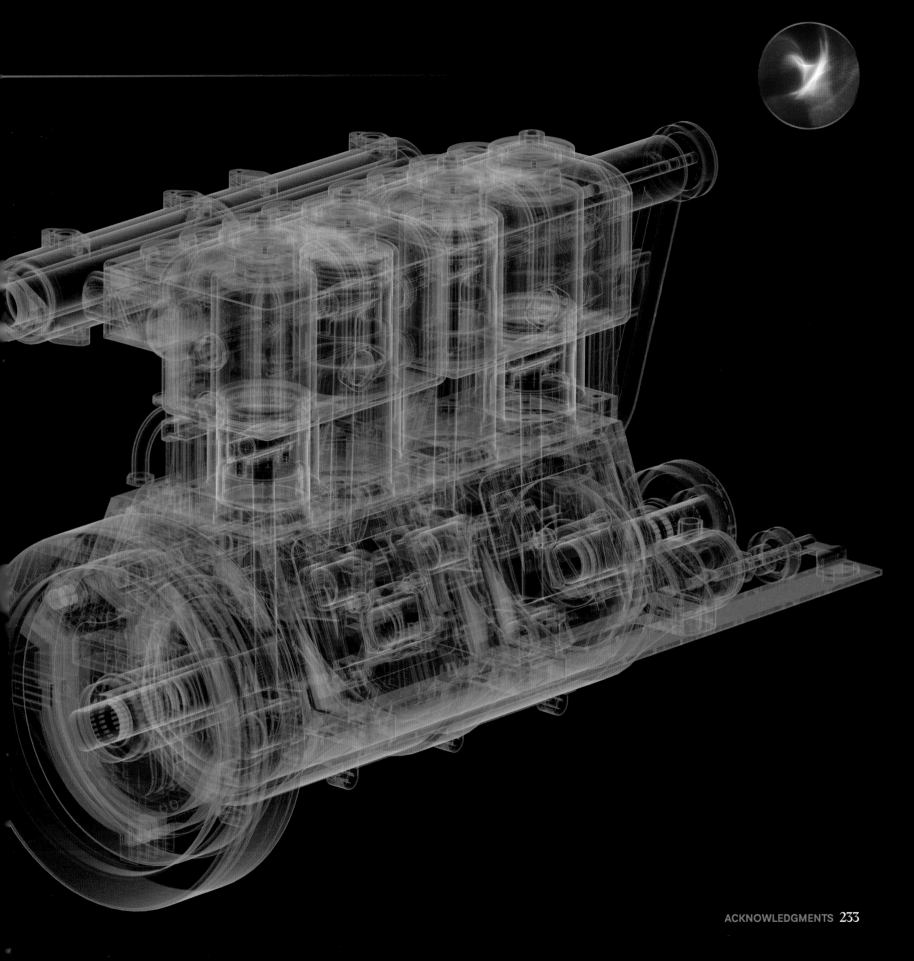

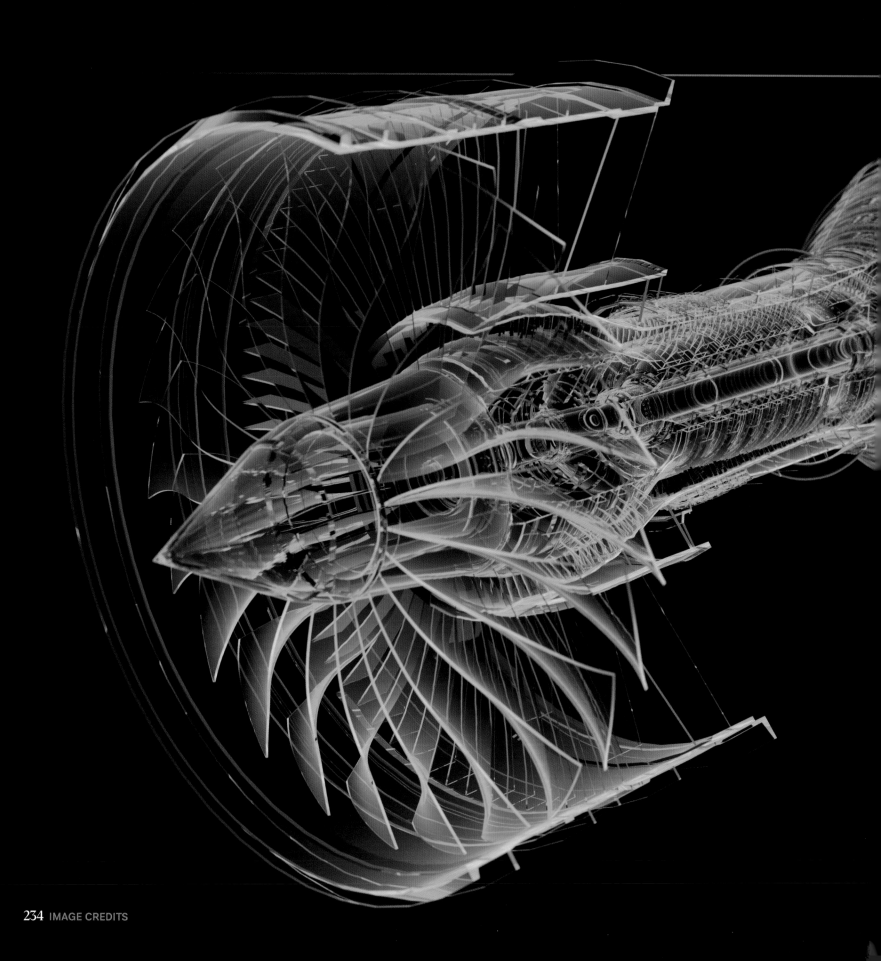

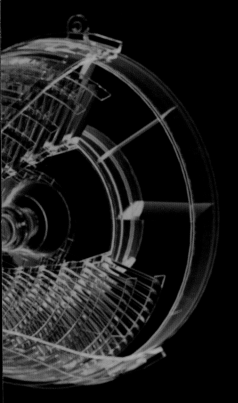

Image Credits

Index